Agency Theory and Executive Pay

Alexander Pepper

Agency Theory and Executive Pay

The Remuneration Committee's Dilemma

Alexander Pepper
Department of Management
London School of Economics and Political Science
London, UK

ISBN 978-3-319-99968-5   ISBN 978-3-319-99969-2   (eBook)
https://doi.org/10.1007/978-3-319-99969-2

Library of Congress Control Number: 2018959560

This Palgrave Pivot imprint is published by the registered company Springer Nature
Switzerland AG
The registered company address is: Gewerbestrasse 11, 6330 Cham, Switzerland

PREFACE AND ACKNOWLEDGEMENTS

A number of people at LSE and elsewhere have been my discussion partners during the last three years while this book has been researched. You know who you are. Thanks to all of you for your many insights. My thanks also to a number of anonymous reviewers who have read all or part of the book at various stages in its development.

I presented an earlier version of Chap. 2 entitled "Rethinking the form, governance and legal constitution of corporations" at the EURAM 2017 conference in Glasgow, Scotland, in June 2017. I am grateful for the helpful comments provided by participants at that workshop.

Zak Yang, an MSc student at LSE, was my research assistant during the summer vacation of 2016 and gathered much of the data used in Chap. 4.

Chapter 5 is based in part upon my article entitled "Applying economic psychology to the problem of executive compensation" published in 2017 in *The Psychologist-Manager Journal*, 20 (4), pp. 195–207. Other material is drawn from Chapter 3 of my book *The Economic Psychology of Incentives* published in 2015 by Palgrave Macmillan.

London, UK   Alexander Pepper

CONTENTS

LIST OF FIGURES

List of Tables

CHAPTER 1

Agency Costs, Coordination Problems, and the Remuneration Committee's Dilemma

Abstract This chapter provides a context for the rest of the book, explaining what is meant by the problem of executive pay, how agency theory has contributed to the problem rather than solved it, and how the critique of agency theory set out in the following chapters might help to solve the problem.

Keywords Theory of the firm • Agency theory • Executive pay

INTRODUCTION

Michael Jensen and William Meckling began their famous article, "Theory of the firm: managerial behavior, agency costs, and ownership structure", which was published in the *Journal of Financial Economics* in 1976, with a quotation from Adam Smith's *The Wealth of Nations*:

> The directors of such companies, however, being the managers rather of other people's money than of their own, it cannot well be expected that they should watch over it with the same anxious vigilance with which the partners in a private co-partnery frequently watch over their own. Like the stewards of a rich man, they are apt to consider attention to small matters as not for their master's honour, and very easily give themselves a dispensation from having it. Negligence and profusion, therefore, must always prevail, more or less, in the management of the affairs of such a company.
>
> Adam Smith (1776) *An Inquiry into the Nature and Causes of the Wealth of Nations.* Book V, Chapter 1, Part III

A. Pepper, *Agency Theory and Executive Pay*,
https://doi.org/10.1007/978-3-319-99969-2_1

1

This much-quoted paragraph appears towards the end of Smith's book in a chapter entitled *On the expenses of public works and public institutions*, which discusses a series of topics that modern economists still wrestle with: the provision of public goods such as roads, bridges, canals, and harbours; collective action problems, where the costs of actions which benefit many fall disproportionately on a few; monopolies—which should be permitted, which discouraged, and how they should be regulated; and agency problems in public corporations, where costs arise because of the different interests of stockholders and managers. Two of these topics—agency problems and collective action—lie at the heart of this short book.

In the language of modern economic theory, agency costs arise when one or more person(s), the principal(s), engage(s) another person or persons, the agent(s), to perform some activity on their behalf, such that decision-making authority is substantially delegated by the principal to the agent. If both persons are utility maximisers, then there is good reason to believe that the agent will not always act in the interests of the principal, resulting in costs—agency costs—which are typically borne by the principal. A specific example of a principal-agent relationship, according to modern economists, is the contractual arrangement between the shareholders and managers of a public corporation.[1]

Adam Smith argued that because the managers of a public corporation do not have the same proprietorial interests as (active) partners in a (trading) partnership they could not be expected to exercise the same level of care and attention in their management of the enterprise. The inevitable result, he concluded, is "negligence and profusion" or, in other words, ineffectiveness and inefficiency. He goes on to argue that joint-stock companies have seldom succeeded without "excessive privilege", such as monopoly trading rights, and he suggests that, even when granted such excessive privilege, they have often mismanaged their enterprises. Adolf Berle (a lawyer and legal scholar) and Gardiner Means (an economist and Berle's one-time research assistant) reached a similar conclusion in their seminal text *The Modern Corporation and Private Property*,[2] which examines the nature of ownership and control of large corporations in the

---

[1] Jensen, M., & Meckling, W. (1976). Theory of the firm: Managerial behavior, agency costs and ownership structure. *Journal of Financial Economics*, 3 (4), pp. 305–360.

[2] Berle, A., & Means, G. (1932). *The Modern Corporation and Private Property*. New York: Macmillan.

United States in the 1930s. They argued that the dispersal of shareholdings in public corporations fundamentally undermined the unity of property rights. Small shareholders holding only fractional property rights over corporations had little incentive or ability to influence the day-to-day management of a company or to hold the managers accountable. Berle and Means identified three different types of relationship comprised in any enterprise: (1) "having an interest in" (in the sense of "Person X has an interest in Enterprise E", i.e., some kind of legal property right); (2) "having power over" ("X has power over E", i.e., de facto possession and control, in the sense of "possession being nine-tenths of the law"); and (3) "acting with respect to" (in other words, "managing", in the sense of "X has the right to manage the day-to-day activities of E"). They go on to describe the evolution of the modern corporation in North America in the following terms. Before the industrial revolution, (1), (2), and (3) were combined and held by the same person or persons. In the nineteenth century, (3) became separated from (1) and (2) with the rise of the professional manager in, for example, the railroad, oil, and steel industries; however, legal and de facto ownership, that is, (1) and (2), remained firmly in the hands of the industrial barons—the Vanderbilts, Rockefellers, Carnegies, and others. In the twentieth century the dispersion of stock ownership over ever-greater numbers of stockholders caused "interest in", that is, (1) to become separated from "power over", that is, (2), so that stockholders became, in the words of Berle and Means, "owners without appreciable control". This power vacuum encouraged managers to exercise greater influence over the enterprises that they managed, described by Berle and Means as "control without appreciable ownership".[3]

It might be said that *The Modern Corporation and Private Property* is long on analysis of the problems of dispersed ownership but relatively short on possible recommendations. Berle and Means do describe (remembering again that they were writing in the early 1930s) three possible futures. First, the traditional logic of property rights, whereby corporations "belong" to their shareholders, might be substantially reinforced, such that managers controlling corporations are placed explicitly in the position of trustees who are required to operate the corporation for the sole benefit of shareholders; although Berle and Means are silent on the point, this would presumably require corporate law and securities

---

[3] Both quotations in this paragraph are from Berle and Means (1932) p. 121.

regulation to be tightened to make these objectives specific. Alternatively, the inexorable logic of laissez-faire economics and pursuit of the profit motive might lead to "drastic conclusions":

> If, by reason of these new relationships, the men in control of a corporation can operate it in their own interests, and can divert a portion of the asset fund of income stream to their own uses, such is their privilege. Under this view, since the new powers have been acquired on a quasi-contractual basis, the security holders have agreed in advance to any losses which they may suffer by reason of such use.[4]

To put it another way, if shareholders' reasonable expectations are satisfied in terms of (1) receiving regular dividends and (2) having the ability to release the value of their shares at any time by selling them on a stock market, then the rent-seeking activities of managers should be regarded as an inevitable and acceptable cost of investing in company shares. This hardly seems a desirable conclusion.

However, Berle and Means do also briefly set out a third possibility, which is frequently overlooked.[5] This is that public corporations could be run in the interests of society as a whole, rather than primarily in the interests of shareholders and managers:

> When a convincing system of community obligations is worked out and is generally accepted, in that moment the passive property right of today must yield before the larger interests of society. Should the corporate leaders, for example, set forth a program comprising fair wages, security to employees, reasonable service to their public, and stabilization of business, all of which would divert a portion of profits from the owners of passive property, and should the community generally accept such a scheme as a logical and human solution of industrial difficulties, the interests of passive property owners would have to give way. Courts would almost of necessity be forced to recognize the result, justifying it by whatever of the many legal theories they might choose. It is conceivable, – indeed it seems almost essential if the corporate system is to survive, – that the "control" of the great corporations should develop into a purely neutral technocracy, balancing a variety of claims by various groups in the community and assigning to each a portion

---

[4] Berle and Means (1932) p. 354.
[5] Bratton, W., & Wachter, M. (2010). Tracking Berle's footsteps: the trail of the Modern Corporation's last chapter. *Seattle University Law Review, 33*(4), pp. 849–875.

of the income stream on the basis of public policy rather than private cupidity.[6]

Sadly, Berle and Means do not really develop their thesis; in particular, they do not explain how it might come about. But I shall return to this final and largely forgotten chapter of *The Modern Corporation* later in this book as I discuss how agency theory might be repaired and corporate governance systems redesigned for the twenty-first century.

The problem of executive pay is central to this book. By "the problem" I mean: (1) pay inflation; (2) the extent to which executive pay is contributing to growing inequality; and (3) public disquiet about how much public corporations pay their top executives. I regard executive pay as a "presenting problem", in the sense that it is the immediate, highly visible, issue of the moment with current corporate governance practice, but it is not the only matter of concern. Commentators have pointed out other problems with contemporary Western corporate governance policy and practice, including short-termism (the way that financial markets force the hands of corporate managers by emphasising the importance of short-term gains over long-term benefits) and sustainability (the apparent inability of corporate governance systems to deal with negative externalities such as pollution and climate change, and the lack of focus generally on corporate social responsibility).[7]

Some people will argue that the level of executive pay is market driven and has nothing to do with agency theory or business ethics. This is not a satisfactory argument for a number of reasons. According to standard microeconomic theory, an efficient market requires many buyers and

---

[6] Berle and Means (1932) p. 356.

[7] The economist John Kay conducted a review of the long-term performance and governance of UK quoted companies at the request Secretary of State for Business, Innovation and Skills between June 2011 and July 2012, leading to the publication of the Kay report (Kay 2012). Professor Colin Mayer's book, *Firm Commitment: Why the Corporation is Failing Us and How to Restore Trust in it*, published in 2013, covers similar ground. More recently, The Purposeful Company Task Force, a consortium of FTSE companies, investment companies, business schools, business consultancies, and policy makers, sponsored by the Big Innovation Centre, has investigated how corporate governance and the capital markets environment in the UK could be enhanced to support the development of value-generating companies, acting with a purpose for the long-term benefit of all stakeholders (Chapman et al. 2017). The Purposeful Company Task Force has also produced a report specifically on the subject of executive remuneration.

sellers, homogeneous products (or at least good substitutes), free market entry and exit, plentiful information, and little economic friction (any factors that inhibit the free operation of the market). The trouble with the market for senior executives is that practically none of these conditions holds good. At any one time only a few top jobs may be open, and only a limited number of suitable candidates may be available. No two senior executives are the same and information about them is far from perfect. Information about prices (what executives are paid) is far from perfect too, despite the best endeavours of governments and regulators in recent years. Finally, all sorts of legal, tax, and accounting factors have an impact on the way senior executives are paid and the types of contracts companies chose to enter into with them. So the standard theory of supply and demand has only a limited amount to say that is helpful about the question of executive pay. Senior executive pay is, on the face of it, certainly at the top end, an example of a "market failure".[8] The labour market for very senior executives does not work efficiently.

It is a curious feature of free markets that in many trades, professions, and occupations the few at the top often earn many times more than the average; in statistical terms, the arithmetic mean (the total of everyone's earnings divided by the number of people in the relevant category) is significantly higher than the median (the mid point in terms of ranking). This phenomenon, which two American economists have called "winner-takes-all",[9] is most noticeable in professional sport. Why do top rank football, basketball, or baseball players earn more in a week than the majority of similar professionals earn in a year? If all the players of a particular sport earned modest amounts, then the coach of one team might reason as follows. If he was able to pay well above the odds, then he would be able to recruit the best players, win lots of trophies, attract large crowds, and secure the best sponsorship deals. The trouble is that other coaches will reason in the same way so that paying high wages becomes a dominant strategy. This is an example of a prisoner's dilemma, where the inevitable logic of the situation leads to a position that is suboptimal for every team;

[8] Shorter, G., & Labonte, M. (2007). The economics of corporate executive pay. *DigitalCommons@ILR*. Retrieved from digitalcommons.ilr.cornell.edu

[9] Frank, R., & Cook, P. (1985). *The Winner-Takes-All Society: How More and More Americans Compete for Ever Fewer and Bigger Prizes, Encouraging Economic Waste, Income Inequality, and an Impoverished Cultural Life*. Free Press.

no team wants to be left with the least able players, thereby running the risk of failure on the field or court with all its consequential financial implications. So everybody decides to pay over the odds.

William Poundstone, in his book the *Prisoner's Dilemma*,[10] describes game theory's most famous puzzle like this. Suppose that two criminals are arrested and imprisoned, with no means of conferring with each other. The police recognise that they do not have enough evidence to convict the pair on the principal charge, but would expect to get both sentenced to a year in prison on a lesser count. Simultaneously, they offer each prisoner a Faustian bargain. They tell each one that if he testifies against his partner in the crime he will go free, while his partner will get five years in prison on the main charge. If both testify against each other, they will both serve two years in jail. Both know that they are being offered the same deal, but they will not learn what the other has decided until they have made their decisions. The two prisoners, who are interested only in their own welfare, rationalise the situation like this. The best result is obtained by testifying against their partner, as long as the other prisoner does not do likewise. But even if he does, the result—a two-year jail sentence—is still better than the worst-case scenario, a five-year sentence. So the rational response for the two prisoners is to testify against each other, even though a better result for both—in aggregate terms—might have been obtained by remaining silent.

Companies face a prisoner's dilemma when it comes to chief executive officers' pay. To demonstrate this, let us assume that all CEOs are paid broadly equal amounts, with only marginal variations in pay justifiable by reference to job size, industry, specialist expertise, and so on. Assume also that in the available population of CEOs 20% are superior to the others and would, if they worked for your company, increase the value of the firm by more than the average. Conversely, 10% are inferior to the others and would, if you employed them, potentially reduce the firm's value. If all companies offered modest remuneration, then it would be in the interests of an individual company to defect and pay over the odds. By doing so, they might attract top talent and (potentially) be more successful than their competitors. Conversely, a company would not want to find itself in the position of paying significantly below average. To do so might mean it could only attract inferior chief executives. No one will congratulate a

---

[10] Poundstone, W. (1992). *Prisoner's Dilemma*. New York: Doubleday.

company's remuneration committee for its financial prudence if the result is a second-rate management team. Thus, offering higher salaries is the dominant strategy, even though by doing so companies will generally be no better off than if they all paid modest salaries. On the other hand, this is better than risking being in the bottom 10%.

We can represent the problem which remuneration committees face (which I call *the Remuneration Committee's Dilemma*) in a pay-off table (see Fig. 1.1).

Scenario 1 is the neutral option; every company pays the market rate and accepts the quality of chief executive they get. In scenario 2, company

Company Y, Z etc.

	Pay market rate	Pay above market
Pay market rate	**Scenario 1** (0, 0)	**Scenario 3** (-10, 0)
Pay above market	**Scenario 2** (+5, 0)	**Scenario 4** (-5,-5)

Company X

+5 = strong preference for (get top performer)

0 = marginal preference for (get satisfactory performer at market rate)

-5 = marginal preference against (get satisfactory performer at above market rate)

-10 = very strong preference against (get inferior performer)

Fig. 1.1 The remuneration committee's dilemma

X defects and pays over the odds in the hope of getting a top performer who will materially influence the value of the company. In scenario 3, company X is left paying the market rate while everyone else pays over the odds, thereby running the risk of hiring inferior talent who will negatively impact the company's net worth. Scenario 4 is the dominant strategy; everyone pays over the odds, but in doing so neither increases nor reduces the likelihood that they will recruit superior talent.

The Remuneration Committee's Dilemma describes a coordination problem—how can remuneration committees achieve a favourable outcome, whereby companies pay a market clearing rate for their top executives, rather than paying over the odds without either increasing the probability of recruiting real stars or of avoiding inferior talent? Shareholders also face a coordination problem, albeit of a different kind. There is a sense in which a public corporation is, to a group of widely dispersed shareholders, a quasi-public good or common pool resource. Minority shareholders in public companies have reasonable expectations of receiving regular dividends and periodic capital gains in excess of safer alternative forms of investment and proportionate to the level of risk they are taking. While agency theory points out the importance of shareholders monitoring the activities of managers, they will wish to do so at minimal cost. They will certainly wish to avoid incurring monitoring costs that materially eat into their income and gains. They would probably be happy to follow any larger shareholder who is prepared to take a lead in monitoring the activities of managers. They may even be prepared to accept rent-seeking behaviour by managers as long as their reasonable expectations of income and gains are met.

Michael Jensen and William Meckling begin their famous article with the quotation from Adam Smith I cited at the start of this chapter because they see the problem identified by Smith as a special case of the economic theory of agency. As Adam Smith and Berle and Means had also pointed out, there are good reasons to believe that the interests of shareholders (principals) and managers (agents) may diverge. However, where Jensen and Meckling depart from the earlier commentators is in their argument that agency theory offers a solution to the problem of divergent interests and the resulting agency cost: "the principal can limit divergences from his interests by establishing appropriate incentives for the agent and by incurring monitoring costs designed to limit the aberrant activities of the agent".[11] They go on to say: "These methods include auditing, formal control systems, budget restrictions, and the establishment of incentive compensation systems which serve to more closely identify the manager's interests with those of outside equity holders".[12] Current theories of corporate governance and executive compensation, and the policies and practices that have developed in Western economies since the publication of Jensen and Meckling's seminal article, are substantially based upon these two ideas.

The main thesis of this book is that Jensen and Meckling were essentially right in their analysis of the agency problem in public corporations but wrong about the proposed solutions. In particular, (1) collective action problems mean that, in a world of self-interested principals with small percentage shareholdings, corporate governance mechanisms are unlikely to achieve all the desired objectives; (2) in practice, incentive contracts have become part of the agency problem rather than a solution and that high-powered incentive contracts give rise to agency costs which lie at the root of many of the current controversies about executive pay; (3) public corporations are real entities, located in time, with unique identities, and individual organisational cultures akin to corporate personalities—they are not, as Jensen and Meckling assert, merely "legal fictions".[13] A deeper analysis of what is wrong with agency theory and the problems that must be repaired in any new theory are the subjects of Chap. 2.

Later in their article on the theory of the firm, Jensen and Meckling argue that organisations are "legal fictions which serve as a nexus for a set of contracting relationships among individuals". A firm is "a focus for a complex process in which the conflicting objectives of individuals (...) are brought into equilibrium with a framework of contracting relations".[14] The nexus of contracts theory of corporations has been taken up enthusiastically by many economists and even by some legal theorists, especially those operating in the "law and economics" tradition.[15] However, this eviscerated view of corporations is, I believe, profoundly wrong. Other legal scholars, philosophers, and some economists have posed a number of

---

[11] Jensen and Meckling (1976) p. 308.

[12] Jensen and Meckling (1976) p. 323.

[13] Jensen and Meckling (1976) p. 310.

[14] Jensen and Meckling (1976) pp. 310–311.

[15] The law and economics tradition is discussed in more detail in Chap. 3.

fundamental questions about the nexus of contracts conception of the firm. Is it a model, a metaphor, or a statement about the "real" nature of companies? To what extent does corporate law determine the nature of the corporation? What does it mean to say that a company has a separate legal personality? How can a firm's identity continue over time notwithstanding changes in its shareholders, managers, key employees, major assets, and, even, in extremis, its whole business.[16] Is a firm a collection of phenomena, an aggregate of individuals, a structure of relations between individuals and assets, or something else? What is the real nature of the corporation? These are ontological questions, philosophical puzzles about the nature of being, becoming, existence, and reality. The ontological status of public corporations (i.e., whether they are real entities or legal fictions) is the subject of Chap. 3. I will argue that in order to answer questions about ownership, participation, and purpose—topics that lie at the very heart of corporate governance—we must first attempt to answer the ontological question about what a public corporation really is. My conclusion, that public corporations are real entities, has important implications for the answers to two questions: (1) who are senior executive agents of; and (2) do public corporations have social, legal, and ethical responsibilities separate from those of their directors and officers?

The coordination problem faced by public corporations is of a type that a community faces when trying to decide how to finance an asset (a public good) which benefits everyone to some extent, but from which no one (including, in particular, free-riding non-contributors) can be excluded. Similar problems arise with common pool resources, like fisheries and forests, which require coordinated action by a consortium of community members to ensure protection for future use, in circumstances where an individual rent-seeking community member might seek to exploit the resource for their own selfish ends by breaking the rules of the consortium. These collective action problems faced by shareholders are examined in more detail in Chap. 4.

---

[16] An interesting case study is the Finnish company, Nokia, which began life in the nineteenth century as a forestry products company, moved into power generation, cable manufacturing, and rubber products, before radically altering its business strategy in the 1980s to focus on mobile telephony. By the end of the twentieth century it had become one of the world's leading mobile telephone handset manufacturers, before going into rapid decline in the last decade.

One of the consequences of the dominant impact of agency theory on business thinking in the latter part of the twentieth century has been the focus on highly elaborate share-based incentive plans, often of baroque complexity, of which the philosopher Joseph Heath, in an article entitled, "The uses and abuses of agency theory", says: "an enormous amount of time and energy has been frittered away designing increasingly clever incentives schemes, to the neglect of more obvious strategies for securing employee loyalty and dedication".[17] One of the problems with long-term incentive plans is that they have not taken account of new thinking in the behavioural sciences. In Chap. 5, I examine the data and explain how psychological factors affecting the perception of risk, uncertainty, complexity, and time cause senior executives to undervalue long-term incentives, thereby contributing to inflation in executive pay. This chapter also examines the difference between extrinsic and intrinsic motivation and explains how intrinsic motivation can be undermined by extrinsic rewards.

Notwithstanding the somewhat critical perspective of many parts of this book, agency theory has many strengths; it would not have dominated scholarly thinking about executive compensation for the last 30 years if it did not. The Oxford philosopher G.E. Cohen talks of his great admiration for John Rawls, before attacking Rawls' famous work *A Theory of Justice* in his own book *Rescuing Justice and Equity.*[18] I have similar admiration for Michael Jensen and other well-known agency theorists. The principal-agent model is, I believe, a good example of a theory which illustrates the thesis that social science theorising is the best thought of from a "model-theoretic" perspective, which recognises the contribution of scholars building on theories in such a way that knowledge accumulates, in contrast to the "law-statement" perspective which dominates the natural sciences, where empirical research supports or refutes general axioms of theory.[19] Accordingly, in Chaps. 5 and 6, I attempt to repair agency theory

---

[17] Heath, J., (2014) "The Uses and Abuses of Agency Theory", in *Morality, Competition, and the Firm – The Market Failures Approach to Business Ethics,* New York, Oxford University Press, Chapter 10, p. 283.

[18] Cohen describes Rawls's *Theory of Justice* (1971) as one of the three most important books in Western political philosophy, comparable with Plato's *Republic* and Hobbes's *Leviathan,* in a section entitled "The Greatness of John Rawls", before starting his extensive critique in *Rescuing Justice and Equality* (2008) pp. 11–14.

[19] Harris, J., Johnson, S., & Souder, D. (2013). Model-theoretic knowledge:the case of agency theory and incentive alignment. *Academy of Management Review, 38*(3), pp. 442–454.

by building into the standard model more realistic assumptions about human behaviour and a better understanding of relevant social norms and legal institutions—a development of the standard framework which has become known as "behavioural agency theory".[20]

THE PURPOSE OF THIS BOOK

The purpose of this book is to critique and repair agency theory is so far as it applies to the relationship between shareholders and executives in public corporations. Some may ask: "why bother – agency theory as a component of the theory of the firm is fatally flawed and should therefore be rejected – far better to develop entirely new theory". That is not my view, as will be apparent. While agency theory has become strongly associated with shareholders and executives, as originally conceived it is of wider application.[21] The principal-agent model is relevant whenever one person contracts with another to perform some activity in circumstances where there is moral hazard (risks taken by the agent are borne by the principal), asymmetric information (the two parties know different things relevant to the contract), and the possibility of adverse selection (rigged trades). Its diagnosis of an economic problem, that costs arise because of the different interests of principals and agents, is entirely sound; it is the proposed solutions to the agency problem when it comes to public corporations that have been found wanting. By repairing agency theory in its application to public corporations my hope is that better solutions to the agency problem will be forthcoming. That is the ultimate objective of this short book.

The next chapter prepares the ground—it explains the important role played by the agency theory in the standard economic model of the firm and identifies some of the problems with this model.

Further Reading
A number of the main themes of this book are covered in a collection of essays by the economically-minded philosopher, Joseph Heath, entitled:

---

[20] See Bosse, D., & Philips, R. (2016). Agency theory and bounded self-interest. *Academy of Management Review*, 41 (2), pp. 276–297, who similarly advocate a "repair" rather than a "replace" strategy.
[21] Ross, S. (1973). The economic theory of agency: the principal's problem. *American Economic Review*, 63(2), p. 134–139.

Morality, Competition, and the Firm, published by Oxford University Press in 2004.

References

Berle, A., & Means, G. (1932). *The Modern Corporation and Private Property.* New York: Macmillan.

Bosse, D., & Philips, R. (2016). Agency Theory and Bounded Self-Interest. *Academy of Management Review, 41*(2), 276–297.

Bratton, W., & Wachter, M. (2010). Tracking Berle's Footsteps: The Trail of the Modern Corporation's Last Shapter. *Seattle University Law Review, 33*(4), 849–875.

Chapman, C., Edmans, A., Gosling, T., Hutton, W., & Mayer, C. (2017). *The Purposeful Company Executive Remuneration Report.* Retrieved from London: http://biginnovationcentre.com/media/uploads/pdf/TPC_ExecutiveRemu nerationReport_26Feb.pdf

Frank, R., & Cook, P. (1985). *The Winner-Takes-All Society: How More and More Americans Compete for Ever Fewer and Bigger Prizes, Encouraging Economic Waste, Income Inequality, and an Impoverished Cultural Life.* New York: Free Press.

Jensen, M., & Meckling, W. (1976). Theory of the Firm: Managerial Behavior, Agency Costs and Ownership Structure. *Journal of Financial Economics, 3*(4), 305–360.

Kay, J. (2012). *The Kay Review of UK Equity Markets and Long-term Decision Making. Final Report.* Retrieved from London: https://www.gov.uk/govern ment/uploads/system/uploads/attachment_data/file/253454/bis-12-917- kay-review-of-equity-markets-final-report.pdf

Poundstone, W. (1992). *Prisoner's Dilemma.* New York: Doubleday.

Ross, S. (1973). The Economic Theory of Agency: The Principal's Problem. *American Economic Review, 63*(2), 134–139.

Shorter, G., & Labonte, M. (2007). *The Economics of Corporate Executive Pay. DigitalCommons@ILR.* Retrieved from digitialcommons.ilr.cornell.edu website.

CHAPTER 2

What's Wrong With Agency Theory?

Abstract This chapter begins by describing the standard model of the firm in organisational economics. It continues by providing a critique of the main premises on which the standard model is based: that shareholders own firms and directors are their agents; that agency costs arise at the level of the firm because of the different interests of shareholders and managers; that man is rational, self-interested, and rent-seeking and there is no non-pecuniary agent motivation. A case study of AstraZeneca is used to illustrate some of the points.

Keywords Theory of the firm • Agency theory • Shareholder primacy • Stakeholder theory

INTRODUCTION

Just as particle physics has a standard model of electromagnetic, strong, and weak nuclear forces and the subatomic particles which they act upon, in a similar way, organisational economics has a standard model of principal and agent relationships which helps to explain the nature of the firm. According to the standard model, firms are "legal fictions" that serve as "a

© The Author(s) 2019   15
A. Pepper, *Agency Theory and Executive Pay*,
https://doi.org/10.1007/978-3-319-99969-2_2

nexus of contracts among individuals"[1] which exist primarily for three reasons; first, in order to save transaction costs in circumstances where (external) market transaction costs exceed the equivalent (internal) governance costs (while there are many references, the locus classicus is the seminal essay by leading institutional economist, Ronald Coase, published in 1937); secondly, to facilitate joint production where team surpluses could not otherwise be allocated among independent subcontractors;[2] thirdly, to provide a vehicle for defining property rights and solving contracting problems.[3] The standard model assumes that shareholders appoint professional managers to make both strategic and everyday tactical decisions on their behalf. The separation of ownership and control in this way creates agency costs as a result of information asymmetry and because self-interested managers do not necessarily act in the interests of shareholders. The agency problem is solved by monitoring (corporate governance) and by constructing high-powered incentive contracts for managers. Incentive contracts are designed to align the interests of shareholders and managers.[4] Corporate governance structures are determined by the principle of "shareholder primacy":[5] shareholders, as residuary beneficiaries, are the firm's ultimate owners; the overriding objective of company managers is to maximise shareholder value.[6]

In the same way that the sociologist Mark Granovetter has argued that neoclassical economics operates with an under-socialised conception of human action,[7] in this chapter I argue that the standard model of managerial agency is "under-institutionalised", in the sense that its assumptions about managerial behaviour, social norms, and legal institutions are oversimplified or simply wrong. I critique the standard model of agency from

[1] Jensen, M., & Meckling, W. (1976). Theory of the firm: Managerial behavior, agency costs and ownership structure. *Journal of Financial Economics,* 3 (4) p. 310.

[2] Alchian, A., & Demsetz, H. (1972). Production, information costs and economic organization. *American Economic Review,* 62 (5) pp. 777–795.

[3] Grossman, S., & Hart, O. (1983). An analysis of the principal-agent problem. *Econometrica,* 51 (1) pp. 7–45.

[4] Jensen and Meckling (1976) p. 308.

[5] Milgrom, P., & Roberts, J. (1992). *Economics, Organisation and Management* (2nd ed.). New Jersey: Prentice-Hall Inc.

[6] Friedman, M. (1970). The social responsibility of business is to increase its profits. *The New York Times Magazine, September 13, 1970.*

[7] Granovetter, M. (1985). Economic action and social structure: the problem of embeddedness. *American Journal of Sociology,* 91 (3) pp. 481–510.

an institutional economics perspective, drawing in particular on the work of legal scholars given the particular importance of corporate law when it comes to the nature of the firm.[8] I argue that standard agency theory's diagnosis of the agency problem in public corporations is essentially correct, especially as it applies to large listed companies in the US and the UK. I also argue that the deductive logic of standard agency theory is valid. However, I contend that some of the major premises on which the standard model is based are wrong and, because the major premises are wrong, the proposed solutions are wrong. Agency theory is a good example of the thesis that social science theorising is better thought of from a "model-theoretic" perspective, which recognises the contribution of scholars building on theories in a such way that knowledge accumulates, in contrast to the "law-statement" perspective which dominates the natural sciences, where empirical research supports or refutes general axioms derived from theory.[9] Accordingly, I propose to build on standard agency theory by embedding within the standard model a better understanding of relevant social norms and legal institutions, adopting a "repair" rather than "replace" strategy.[10]

A number of the arguments advanced in this chapter are consistent with earlier pronouncements by management scholars about stewardship theory.[11] Stewardship theory holds that there is no inherent, general problem

---

[8] Hodgson, G. (2015). *Conceptualizing Capitalism: Institutions, Evolution, Future.* Chicago, Ill: University of Chicago Press; Orts, E. (2013). *Business persons: A Legal Theory of the Firm.* Oxford: Oxford University Press.

[9] Harris, J., Johnson, S., & Souder, D. (2013). Model-theoretic knowledge: the case of agency theory and incentive alignment. *Academy of Management Review,* 38 (3) pp. 442–454. Oliver Williamson quoting Allen Newell, puts it like this: "New theories rarely appear full blown but evolve through a progression during which the theory and the evidence are interactive – 'theories cumulate. They are refined and reformulated, corrected and expanded. Thus, we are not living in the world of Popper...Theories are not shot down with a falsification bullet...Theories are more like graduate students – once admitted you try hard to avoid flunking them out...Theories are things to be nurtured and changed and built up.'" Williamson, O. (2011). Corporate governance: a contractual and organizational perspective. In L. Sacconi, M. Blair, R. Freeman, & A. Vercelli (Eds.), *Corporate Social Responsibility and Corporate Governance.* Palgrave Macmillan., p. 5.

[10] Bosse, D., & Philips, R. (2016). Agency theory and bounded self-interest. *Academy of Management Review,* 41 (2) pp. 276–297.

[11] Donaldson, L., & Davis, J. (1991). Stewardship theory or agency theory: CEO governance and shareholder returns. *Australian Journal of Management,* 16 (1) pp. 49–64. Davis, J., Schoorman, F., & Donaldson, L. (1997). Toward a stewardship theory of management. *Academy of Management Review,* 22 (1) pp. 20–47.

of executive motivation and behaviour. Managers are regarded as stewards whose motives are closely aligned with the objectives of their principals. Agency theory and stewardship theory are regarded either as opposing models or as contingent theories whose fit is dependent upon the particular organisational context. The ontological argument, which is set out in this paper, differs from stewardship theory in the way that it attempts to repair agency theory, rather than offering a competing model. By excluding factors such as identity, power, managerial philosophy, and culture from the analysis, it is also far more parsimonious than stewardship theory. My argument is that a solid base for a revised theory of managerial agency in large corporations can be constructed by replacing a number of standard agency theory's current premises with new, more realistic assumptions about the behaviour of agents and principals.

This chapter proceeds by briefly describing the standard model of agency in firms, then by setting out a critique which draws on the work of scholars writing in the institutional economics tradition. It examines three of the major premises on which agency theory is based, explains why these assumptions are flawed, proposes revised premises, and deduces from these revised premises how agency theory can be repaired. It concludes by proposing various ways in which the standard proposed policy solutions to the agency problem should be revised.

THE STANDARD MODEL OF THE FIRM IN ORGANISATIONAL ECONOMICS

The general principal-agent model focuses on bilateral arrangements where a principal (conventionally "her") hires an agent (conventionally "him") to carry out some activity on her behalf.[12] In its more specific application to companies, agency theory postulates, inter alia, that in order to motivate managers (agents) to carry out actions and select effort levels that are in the best interests of shareholders (principals), boards of directors, acting on behalf of shareholders, must design incentive contracts which make an agent's compensation contingent on measurable

---

[12] Ross, S. (1973). The economic theory of agency: the principal's problem. *American Economic Review*, 63 (2) pp. 134–139; Spence, M., & Zeckhauser, R. (1971). Insurance, information and individual action. *American Economic Review*, 61 (2) pp. 380–38.

performance outcomes.[13] The model is underpinned by two propositions generally attributed to Milton Friedman: first, that corporate executives are employed by the owners of a business, that the owners of a business are the shareholders, and that the sole responsibility of a business is to increase its profits;[14] secondly, that it does not matter how realistic or unrealistic the behavioural assumptions of a social scientific theory are as long as the theory's predictions are accurate.[15]

Criticisms of agency theory in its application to public corporations have been advanced by both empiricists and theoreticians. The first problem for agency theorists is that empirical evidence gathered over the past 35 years has failed to establish a statistically significant link between executive pay and stock price performance, as predicted by agency theory. In 1990 Michael Jensen and Kevin Murphy were unable to find a statistically significant connection between CEO pay and performance.[16] Ten years later Tosi, Werner, Katz, and Gomez concluded that incentive alignment as an explanatory agency construct for CEO pay was, at best, weakly supported by the evidence, based on their meta-analysis of over 100 empirical studies.[17] A review by Carola Frydman and Raven Saks of the US executive

[13] Jensen and Meckling (1976).

[14] Friedman (1970).

[15] Friedman, M. (1953/2008). The methodology of positive economics. In D. Hausman (Ed.), *The Philosophy of Economics – An Anthology. Third Edition* (pp. 145–178). Cambridge: Cambridge University Press. (Reprinted from: The Philosophy of Economics: An Anthology. Hausman, D. 2008).

[16] Jensen, M., & Murphy, K. (1990). Performance pay and top-management incentives. *Journal of Political Economy,* 98 (2) pp. 225–264. When Jensen and Murphy failed to find a statistically significant connection between CEO pay and performance, they argued that this was the result of political forces at the heart of the corporation and that companies should provide a greater proportion of total compensation in the form of incentive pay, thus switching from a positive to a normative line of argument. I call this the "J- twist". Paul Samuelson (1963) described Milton Friedman's thesis that the truth of the assumptions is irrelevant to the acceptability of a theory, provided that the theory's predictions succeed, as the "F-twist". Steve Keen argues that Tony Lawson provides the "L-correction" to the "F-twist" by forcing economics to consider its ontology – see Lawson (2015) postface and Chap. 3. In the same spirit, one of the aims of this book is to point out and provide a correction to Jensen's "J-twist".

[17] Tosi, H., Werner, S., Katz, J., & Gomez-Mejia, L. (2000). How much does performance matter? A meta-analysis of CEO pay studies. *Journal of Management,* 26 (2) pp. 301–339. Two subsequent meta-analytic reviews have continued to provide evidence that CEO pay and financial performance are not closely related: see van Essen, M., Otten, J., & Carberry,

compensation data covering the period 1936–2005 concluded that neither agency theory nor the managerial power hypothesis was fully consistent with the available evidence.[18] Optimal contracting theorists
(mathematical economists who are the present-day descendants of mainstream agency theorists) now appear to accept that the strongest empirical
correlation is between executive pay and firm size, not between executive
pay and firm performance as predicted by agency theory.[19] Baker, Jensen,
and Murphy have even called this: "the best documented empirical regularity regarding levels of executive compensation".[20]

The major premises of standard agency theory are as follows: first, that
firms are owned by their shareholders and that directors are agents of
shareholders; secondly, that all agency costs arise at the level of the firm;
thirdly, that man is rational, self-interested, and rent-seeking and there is
no non-pecuniary agent motivation. I address these premises, in turn,
below:

First Premise: Shareholders Own Firms and Directors Are Their Agents

In 1970 Milton Friedman wrote

> A corporate executive is an employee of the owners of the business. He has
> direct responsibility to his employers. That responsibility is to conduct the

E. (2015). Assessing managerial power theory: a meta-analytic approach to understanding
the determinants of CEO compensation. *Journal of Management, 26*(2), pp. 164–202, and
Aguinis, H., Gomez-Mejia, L., Martin, G., & Joo, H. (2018). CEO pay is indeed decoupled
from CEO performance: charting a path for the future. *Management Research, 16*(1),
117–136.

[18] Frydman, C., & Saks, R. (2010). Executive compensation: a new view from a long-term
perspective, 1936–2005. *The Review of Financial Studies,* 23(5) pp. 2099–2138. The managerial power hypothesis can be found in Bebchuk, L., & Fried, J. (2004). *Pay without performance – the unfilled promise of executive compensation.* Cambridge, Mass: Harvard University
Press.

[19] See Gabaix, X., & Landier, A. (2008). Why has executive pay increased so much?
Quarterly Journal of Economics, 123 (1) pp. 49–100, and Edmans, A., & Gabaix, X. (2016).
Executive compensation: a modern primer. *Journal of Economic Literature,* 54 (4)
pp. 1232–1287.

[20] Baker, G., Jensen, M., & Murphy, K. (1988). Compensation and incentives: practice vs
theory. *Journal of Finance,* 43 (3) p. 609.

business in accordance with their desires, which generally will be to make as much money as possible, while conforming to the basic rules of the society, both those embodied in law and those embodied in ethical custom.[21]

Here, as elsewhere, Friedman makes the tacit assumption that the "owners of the business" are its shareholders, following standard economic thinking that ownership of corporations is linked to the provision of capital. Other scholars have pointed out that property rights within a firm are not unitary and can be easily disaggregated.[22] Shareholders hold property rights in company shares, entitling them to residual cash flows, including dividends and proceeds from stock buybacks;[23] they also have representation rights, for example, to vote in certain narrowly prescribed circumstances; however, this does not make them a public corporation's "owners"—they do not have a complete bundle of rights which would make them owners in any conventional sense.

One of the main theoretical challenges to the standard model has come from the stakeholder theory of the firm.[24] Stakeholder theory questions agency's theory's central concept of shareholder primacy, arguing instead that shareholders are only one of a number of important interest groups; other stakeholders include employees, customers, suppliers, and local communities. Joseph Heath provides a typology, proposing that there are many different types of stakeholder theory, including ontological stakeholder theory (a theory about the fundamental nature and purpose of the corporation), strategic stakeholder theory (which argues that devoting sufficient resources and managerial attention to stakeholders generally will tend to have positive performance outcomes in terms of profitability, revenue, and share price growth), and corporate law stakeholder theory (which proposes that stakeholder theory more accurately describes the legal nature of corporations than the standard model, and provides insights into how corporate law should be developed to better reflect ontological,

---

[21] Friedman, M. (1970). The social responsibility of business is to increase its profits. *The New York Times Magazine, September 13, 1970.*

[22] Heath, J. (2014). *Morality, Competition, and the Firm.* NY, USA: Oxford University Press.

[23] Ghoshal, S. (2005). Bad management theories are destroying good management practices. *Academy of Management – Learning & Education,* 4 (1) pp. 75–9.

[24] The landmark text is Freeman, R. (1984/2010). *Strategic Management: A Stakeholder Approach.* Cambridge, UK: Cambridge University Press.

Table 2.1 Heath's typology of stakeholder theories

Type of theory	Description
Ontological stakeholder theory	A theory about the fundamental nature and purpose of the corporation
Explanatory stakeholder theory	A positive theory that purports to describe and explain how corporations and managers actually behave in practice
Strategic stakeholder theory	Argues that devoting sufficient resources and managerial attention to stakeholders generally will tend to have positive performance outcomes in terms of profitability, revenue, and share price growth
Branding and culture stakeholder theory	A theory about how a commitment to pay extraordinary attention to the interests of particular stakeholder groups (e.g., customers and employees) can become a fundamental aspect of a firm's branding and corporate culture
Deontic stakeholder theory	An approach which proposes that stakeholder theory helps to determine the rights and duties of stakeholders and managers from an ethical perspective
Managerial stakeholder theory	A catch-all theory of management that helps leaders and managers realise the strategic benefits of stakeholder theory
Governance stakeholder theory	An approach which proposes that stakeholder theory explains how different stakeholder groups should exercise oversight and control over managers
Regulatory stakeholder theory	A theory that defines which interests and rights of specific stakeholder groups ought to be protected by government regulation
Corporate law stakeholder theory	Argues that stakeholder theory more accurately describes the legal nature of corporations than the standard model, and provides insights into how corporate law should be developed to better reflect ontological, deontic, and governance approaches to corporations

deontic, and governance approaches to corporations).[25] Heath's full typology is set out in Table 2.1.

Perhaps the most notable example of corporate law stakeholder theory[26] is the "team production theory of corporate law" advanced by

[25] Heath, J. (2014).

[26] Heath (2014) points out that corporate law varies significantly from country to country and between states in the United States. This presents certain difficulties when attempting to generalise principles drawn from close legal analysis. The implications of this are examined further in Chap. 4.

Margaret Blair (an economist) and Lynn Stout (a legal scholar).[27] They argue that, while agency theory may be important in understanding the private business firm, it does not necessarily provide the same insights into our understanding of public corporations. In a closely held private company stock ownership is often concentrated in the hands of a small number of investors (principals) who select, appoint, and exercise tight control over the board of directors (agents). However, in the case of public corporations, corporate law does not treat directors as the agents of shareholders but as something quite different. They are not charged with serving shareholders' interests alone but with serving the interests of the company. In the eyes of the law, corporate directors are a unique form of fiduciary who more closely resemble trustees than agents. They owe fiduciary duties of loyalty and care to the company, not to shareholders. In the US, legal scholars who subscribe to the standard model rely on the decision in the famous case of Dodge versus Ford Motor Company (Michigan 1919) when the court sided with the Dodge brothers, shareholders who wanted dividends to be maximised, and against Henry Ford, who believed that the company's prosperity should be shared with other stakeholders, including assembly line workers and the local community. That case also affirmed the business judgement rule, which gives corporate executives in the United States wide latitude in how to run a company. However, the primary job of the directors of a public corporation is to act, in effect, as trustees for the corporation itself. They are thus not merely the agents of shareholders, pursuing shareholders' interests at the expense of employees, creditors, and other team members.

Blair and Stout propose that public corporations comprise teams of people making specific investments in the form of both financial and human capital who enter into a complex agreement to work together for mutual gain under a "mediating hierarchy". However, they are careful to distance themselves from other stakeholder theorists who believe that corporate law ought to require directors to serve consumers, creditors, and the public as a whole as well as shareholders and employees. Michael

---

[27] There are many references, most notably Blair and Stout (1999) A team production theory of corporate law. *Virginia Law Review*, 85 (2) pp. 247–328, but also including: Blair (1995, 1996) and Stout (2012). "Team production" is a reference to the economic theory of the firm advanced by Alchian & Demsetz (1972). Blair & Stout also make reference to the conventional theory of corporate agency relationships, described in this book as "the standard model", which they refer to as the "grand-design principal-agency model".

Jensen has in any case identified a fundamental difficulty with these kind of multi-stakeholder theories, pointing out the logical impossibility of maximising in many dimensions at the same time except in unusual circumstances in which all the dimensions are monotonic transformations of one another.[28] For example, if a company's directors have to choose, in the teeth of a recession, between protecting employment, maintaining the company's presence in all the communities in which it currently operates, and paying a dividend, then they are expressing a preference between the utility of employees, the utility of other residents in local communities, and the utility of shareholders. This has been described as the "multi-principal problem" of which it has been said: "a manager told to serve two masters has been freed from both and is answerable to neither".[29] Edward Freeman has suggested that managers must "act like King Solomon" in adjudicating among the claims of various stakeholder groups;[30] the risk is that giving managers such freedom to balance rival claims would create extraordinary agency risks.

Stephen Bainbridge, another legal theorist, has also questioned the idea that shareholders are owners of companies.[31] Shareholders hold property in the form of shares which provide various (limited) rights: to receive dividends (but only if declared by directors); to vote on certain matters of importance, including the appointment and reappointment of directors (but only from a slate of candidates proposed by the board); and to veto major transactions (but only if a coalition of shareholders representing the necessary proportion of total votes prescribed by law can be assembled). They also have the right to receive residual assets in a winding-up, but this rule rarely operates in practice: public companies tend to be wound-up only when bankrupt, when there are often no residual assets. Shareholders are not automatically entitled to enter company premises, to use company assets, or to arrange transactions on the company's behalf; they surely would be if they were genuinely a firm's "owners". According to Antony Honoré, "full ownership" is characterised by 11 "incidents" or indicators:

[28] Jensen, M. (2001). Value maximization, stakeholder theory, and the corporate objective function. *Journal of Applied Corporate Finance*, 14 (3) pp. 8–22.

[29] Heath, J. (2014) p. 62; Easterbrook, F., & Fischel, D. (1991). *The Economic Structure of Corporate Law*. Cambridge, Mass: Harvard University Press, p. 38.

[30] Freeman, R. (1984/2010).

[31] Bainbridge, S. (2003). Director primacy: the means and ends of corporate governance. *Northwestern University Law Review*, 97 (2) pp. 547–606.

the right to possess, the right to use, the right to manage, the right to income, the right to capital, the right to security, the rights of transmissibility and absence of term, the prohibition of harmful use, liability to execution, and incidence of residuarity.[32] Of these indicators, only three (income, capital, and residuarity) appear with any certainty to be met when it comes to shareholders "ownership" of a firm, as opposed to their ownership of shares which convey an interest in the firm.[33]

As an alternative to shareholder primacy and stakeholder theory, Bainbridge proposes "director primacy" which treats the corporation as a vehicle for the board of directors to hire various factors of production. In his model the board is: "a sort of Platonic guardian serving as a nexus for the various contracts comprising the corporation".[34] This approach is consistent with Blair and Stout's analysis that the duties of directors of public corporations more closely resemble those of trustees rather than agents, although Bainbridge departs from Blair and Stout in various other ways, including the question of who directors are trustees for, and especially when it comes to the notion of the board as a mediating hierarchy. Both approaches are broadly consistent with an alternative construction of the overriding responsibility of directors, latterly proposed by Michael Jensen, which he calls "enlightened stakeholder theory" and "total firm value maximization" (TFVM).[35] This postulates that long-term value maximisation of the whole firm (i.e., as distinct from its shareholders) should be the primary objective of company managers. Jensen says: "maxi-

---

[32] Honoré, A. (1961). Ownership. In A. Guest (Ed.), *Oxford Essays in Jurisprudence* (pp. 107–147). Oxford, UK: Oxford University Press.

[33] Also relevant are the five different types of property rights that have been identified (by their presence or absence) in empirical studies of common pool resources systems. These are: *access*, the right to enter a defined physical property; *withdrawal*, the right to draw an income from a common pool resource; *management*, the right to regulate the patterns of use of common pool resources and to transform a resource system by making investments and improvements; *exclusion*, the right to determine who has access and withdrawal rights; and *alienation*, the right to sell or lease any of the other rights – see Poteete, A., Janssen, M., & Olstrom, E. (2010). *Working Together – Collective Action, the Commons, and Multiple Methods in Practice.* Princeton & Oxford: Princeton University Press, p. 95. Shareholders have some (i.e., access, withdrawal, and alienation rights) but not all of these. For more on the relevance of the literature on common pool resources to public corporations, see Chap. 4.

[34] Bainbridge (2003) pp. 550–51.

[35] Jensen (2001).

mising the total market value of the firm – that is the sum of the market values of the equity, debt and any other contingent claims outstanding on the firm – is the objective function that will guide managers in making the optimal trade-offs among multiple constituencies".[36] He continues: "It is a basic principle of enlightened value maximisation that we cannot maximise the long-term market value of an organisation if we ignore or mistreat any important constituency. We cannot create value without good relations with customers, employees, financial backers, suppliers, regulators, and communities". Andrew Keay, an English law scholar, has demonstrated how a watered-down version of enlightened stakeholder theory, which he calls "the enlightened shareholder value principle", applies in UK company law as a result of work undertaken by the Company Law Review Steering Group, which published several reports between 1998 and 2001.[37] After much debate about stakeholder theory and shareholder primacy, the provisions which were eventually incorporated into section 172 of the UK Companies Act 2006 in effect confirmed the prevailing view that companies are run for the benefit of their members (i.e., shareholder primacy) but require directors to have regard to the interests of employees, relationships with suppliers and customers, the impact of a company's operations on communities and the environment, and the desirability of maintaining a reputation for high standards of business conduct. They also require directors to consider the long-term consequence of their decisions and the need to act fairly between the members of the company.

Keay has separately argued that a version of TFVM, which he calls the "entity maximisation and sustainability model" (EMS), should become the corporate objective of British public companies. Two important elements of both the TFVM and EMS ways of answering the question "what is a company for?" should be specifically noted. First, value is to be maximised in the long run, not the short run; it is thus aligned with demands that economic policy and regulation should emphasise the benefits of long-term corporate investment.[38] Secondly, it is firm value, not shareholder value, which is to be maximised. This admits the legitimate interest

---

[36] Jensen (2001) p. 16.
[37] Keay, A. (2013). *The Enlightened Shareholder Value Principle and Corporate Governance.* Abingdon, UK: Routledge.
[38] Kay (2012); Mayer (2013).

in value creation of at least some other stakeholders, including managers and other employees who make long-term personal investments in the company. Just as shareholders contribute financial capital to a firm, so it is alleged that managers and employees contribute human capital, being the sum total of their knowledge, skill, experience, and intelligence; therefore, it is argued that they also have contingent claims outstanding on the firm. Blair and Stout describe these kinds of investments as "specific assets", in much the same way as Oliver Williamson talks about "asset specificity".[39] The TFVM principle can itself be derived from one of the fundamental principles of social welfare; that society's object is to maximise total utility, to create the greatest good for the greatest number, to maximise the efficiency with which society uses resources to create wealth and minimise waste. To quote Jensen again, "moreover, we can be sure...apart from the possibilities of externalities and monopoly power – that using this value criterion will result in making society as well off as it can be".[40]

Second Premise: Agency Costs Arise at the Level of the Firm Because of the Different Interests of Shareholders and Managers

A further issue for the standard model is that the upper echelons of a public corporation involve a number of different principal-agent and fiduciary relationships. In the language of agency theory, shareholders (principals) appoint directors (agents); the directors in turn (now acting in the role of principals) appoint managers as their agents. Agency costs can arise at both levels. Some directors are also managers and therefore wear two hats, sometimes acting as agents (on behalf of shareholders) and sometimes as principals (as members of the board of directors when instructing other managers). This creates the potential for role confusion, although the implications of dual-role conflicts of interest are not examined further here.

To complicate matters, once appointed, directors become fiduciaries, owing their primary duty to the company rather than to its shareholders. They take on an open-ended set of responsibilities to the corporation,

---

[39] Blair and Stout (1999); Williamson, O. (1975). *Markets and Hierarchies*. New York: The Free Press.
[40] Jensen (2001) p. 16.

empowering them to manage the corporation's business and affairs, and imposing upon them a duty of care in exercising those powers. They are expected to exercise a high degree of care, skill, and diligence in carrying out their duties. They are required to show loyalty to the company. They are not in any strict legal sense agents of stockholders.[41] To complicate matters further, many shareholders in public corporations are in fact pooled investment funds, run by investment managers (agents) who invest money on behalf of individuals, firms, and other funds (principals). This creates a further set of agency and fiduciary relationships, and hence further agency costs. Thus, rather than a simple conflated agency relationship between shareholders and managers, as posited by many subscribers to the standard model, it must be recognised that public company governance structures involve a complex web of fiduciary and agency relationships requiring a more sophisticated institutional analysis than allowed for by the mathematical models of many modern agency theorists.

To illustrate these points, consider the case of AstraZeneca plc., a large pharmaceutical company listed on the London Stock Exchange. At the time of writing, AstraZeneca's board comprised two executive directors (the CEO and CFO) and ten non-executive directors, including the chairman and senior independent non-executive director. The company also had a non-statutory management board, known as the "senior executive team", comprising the two executive directors and ten other senior managers. While the main board was said to be responsible for strategy, policy, corporate governance, and monitoring the performance of management, the senior executive team was responsible for management, development, and performance of the business, and for developing a strategy for review by the main board. Both the board and senior executive team were described as being "accountable to shareholders for the responsible conduct of the business and its long-term success". This type of structure, with the management board referred to variously in different companies as the "executive board", "executive committee", "management committee", and so on, has become commonplace in the FTSE 100. One

[41] Clark, R. (1985). Agency costs versus fiduciary duties. In J. Pratt & R. Zeckhauser (Eds.), *Principals and Agents: The Structure of Business*. Boston, MA: Harvard Business School Press.

consequence is that there is no requirement to disclose the pay of senior executives on the management board unless they are also members of the main board.

AstraZeneca's top 100 shareholders controlled nearly 64% of the company's ordinary share capital. Seventy-five of these were institutional investors, managing funds on behalf of individuals and pension funds. The largest investors were Blackrock, with an 8% shareholding, followed by Wellington Management and Capital Group, both owning more than 3%. While Wellington and Capital Group are privately owned investment management companies, Blackrock is itself a public company, listed on the New York Stock Exchange. Its major shareholders included PNC Financial Services Group (a listed investment management company owning 21% of Blackrock), Wellington, Vanguard (another privately owned investment management company), and Norway's sovereign wealth fund. In turn, PNC's major shareholders were Vanguard, Old Mutual, State Street, and Capital Group. Blackrock also owned 4.8% of PNC.

Figure 2.1 maps shareholders owning more than 3% of the share capital of AstraZeneca in December 2015, that is, Capital Group, Wellington (both private investment companies), and Blackrock (a listed investment management company). Major shareholders in Blackrock are also mapped, as, in turn, are the shareholders of its major listed shareholders. The diagram stops at this third level of analysis.

In all this we see a complex web of holdings and cross-holdings, and of agency and fiduciary relationships. Retail investors (principals) make investments in pooled funds. The directors of these funds appoint investment managers as agents to manage investments on their behalf. Both fund directors and investment managers have fiduciary responsibilities. The funds, along with other shareholders (principals), appoint the board of directors of AstraZeneca. The board of directors (acting as agents of shareholders) appoint the CEO. The CEO (now acting as principal) in turn appoints other senior managers (agents of the company). Executive and non-executive directors owe fiduciary responsibilities to the company (but not, strictly speaking, to the company's shareholders).

Agency costs arise at many levels. Because investment firms are sometimes privately held (i.e., private companies or partnerships), transparency (e.g., of compensation costs) is less than for public corporations. Listed investment management companies employ investment executives below

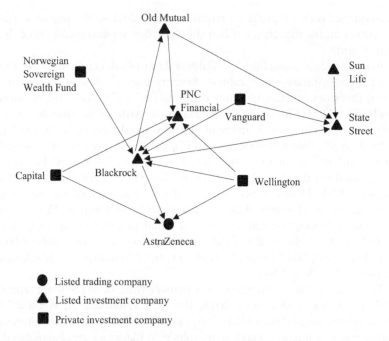

Fig. 2.1 Major shareholders in AstraZeneca – December 2015

board level whose pay is not separately disclosed. It is highly probable that agency costs will arise in pooled investment funds because of the collective action problems which are likely to be extant among widely dispersed retail investors.[42] (I expand on the nature of collective action problems in the next chapter). This is over and above the agency costs which arise in the trading company.

---

[42] This is consistent with the findings of the review carried out by John Kay on behalf of the UK Government's Department of Business, Innovation and Skills in 2012 (Kay 2012).

THIRD PREMISE: MAN IS RATIONAL, SELF-INTERESTED, AND RENT-SEEKING, AND THERE IS NO NON-PECUNIARY AGENT MOTIVATION

An underlying assumption of the standard model is that companies are profit-making, that principals and agents are both rational and rent-seeking, and there is no non-pecuniary agent motivation. Economists recognise that these assumptions are an oversimplification but argue that the most efficient way to construct theory is to adopt a reductive approach. David Hume took a similar line when he wrote in his essay "Of the Independency of Parliament":

> Political writers have established it as a maxim, that, in contriving any system of government, and fixing the several checks and controls of the constitution, every man ought to be considered a *knave*, and to have no other end, in all his actions, than private interest. By this interest we must govern him, and, by means of it, make him, notwithstanding his insatiable avarice and ambition, co-operate to public good...It is, therefore, a just *political* maxim, *that every man must be supposed a knave:* Though, at the same time, it appears somewhat strange, that a maxim should be true in *politics*, which is false in *fact*. But to satisfy us on this head, we may consider, that men are generally more honest in their private than in their public capacity...Honour is a great check upon mankind.[43]

Many scholars now dispute Hume's approach. Julian Le Grand of the London School of Economics has examined the motivation of agents in the public sector in the light of David Hume's famous dictum. Le Grand argues that, both as a matter of fact (i.e., as a positive theory of what is) and in terms of what should be the case (i.e., as a normative theory), the public sector is populated by "knights" as well as "knaves". In Le Grand's terminology, while knaves are "self-interested individuals who are motivated to help others only if by doing so they will serve their private interests", knights are "individuals who are motivated to help others for no private reward, and indeed who may undertake such activities to the

---

[43] Hume, D. (1804). Of the Independency of Parliament *Essays, Moral, Political, and Literary*. Edinburgh: Bell & Bradfute.

detriment of their own private interests".[44] Hume refers to the latter as a "man of honour", and an "honest man". (In Hume's day there was less concern about using terminology that was gender neutral.) To put it another way, the distinction is between, on the one hand, those who are motivated to perform only those activities that are of direct benefit to their own material welfare, such as their own personal consumption of material goods, and, on the other hand, those that are motivated to engage in other directed activities, that is, activities which benefit others and which do not positively affect their own material welfare. Le Grand's argument is that, while both types are found in public service, knightly behaviour is needed if the public sector is to operate efficiently. He also argues, drawing on the work of the Swiss economist Bruno Frey, that increasing in monetary incentives can "crowd-out" intrinsic motivation. Besley and Ghatak have argued that "motivated agents" can be found in public and non-profit organisations, where activities coalesce around a "mission", whose economic behaviour is affected by intrinsic motivation.[45] The economist Samuel Bowles has advanced the same thesis in his book entitled *The Moral Economy – Why Good Incentives are No Substitute for Good Citizens.*[46] If further evidence is needed, Martin Nowak, a mathematical biologist based at Harvard, has published an extensive body of research on human behaviour and evolution. He argues that man should be thought of as a complicated hybrid species ruled as much by emotion and altruism as by reason and selfishness. He demonstrates that human evolution would not have been possible if humans really were relentlessly bent on maximising purely selfish rewards.[47]

Reverting to corporations, corporate governance, and the standard model of principal and agent relationships in firms, the Australian

[44] Le Grand, J. (2003). *Motivation, Agency and Public Policy.* Oxford: Oxford University Press.
[45] Besley, T., & Ghatak, M. (2005). Competition and incentives with motivated agents. *American Economic Review,* 95 (3) pp. 616–636.
[46] Bowles, S. (2016). *The Moral Economy – Why Good Incentives are no Substitute for Good Citizens.* New Haven, CT: Yale University Press.
[47] Nowak, M., & Highfield, R. (2011). *Super Cooperators – Evolution, Altruism and Human Behaviour.* Edinburgh: Canongate. See also Nowak, M., Page, K., & Sigmund, K. (2000). Fairness versus reason in the ultimatum game. *Science,* 289 (5485), pp. 1773–1775, and Sigmund, K., Fehr, E., & Nowak, M. (2001). The economics of fair play. *Scientific American,* 286(1), pp. 82–87.

management scholar Lex Donaldson has pointed out that the characterisation of managers in the standard model is almost entirely negative, and argues that this is deeply unhelpful. If organisational economists regard managers as self-interested, opportunistic agents requiring close monitoring by their principals, as well as needing extrinsic incentives to motivate their every action, this will inevitably influence their policy recommendations.[48] There is even a risk that managers might become what the policy recommendations assume them to be. In a famous essay entitled "Bad management theories are destroying good management practices", Samantra Ghoshal argued that reductive theories such like agency theory cause great damage.[49] No matter how cleverly designed to harness managers' self-interest, incentives alone cannot provide the foundations of good corporate governance. The erosion of ethical, social, and intrinsic motivations essential to good governance could be the unintended consequence of policies, including the excessive use of monetary incentives to guide individual behaviour, which agency theorists favour. The reliance on high-powered incentives to align the interests of shareholders and managers has become another example of what the sociologist Donald Mackenzie describes as "performativity", that economics creates the phenomena it describes.[50] Because economists advocate the use of high-powered incentives to motivate executives and to align their interests with those of shareholders, such incentives have become the norm, even though they have not typically achieved their ultimate objective of improving corporate performance.

A central thesis of this book is that, in corporate governance as in public government, we should recognise the need for ethical behaviour, for honest and honourable men and women who are motivated by mission and regard for others and not just by self-interest. There is a powerful argument that says one part of the logic of the standard model should be reversed, and managers seen not as opportunistic self-regarding agents, but as autonomous intrinsically motivated principals who are central to the enduring mission of the firm.[51] The board of directors and executive

---

[48] Donaldson, L. (1995).

[49] Ghoshal, S. (2005).

[50] Mackenzie, D. (2007). Is economics performative? In D. Mackenzie, F. Muniesa, & S. L (Eds.), *Do Economists Make Markets?* Princeton, New Jersey: Princeton University Press; see also Hodgson 2015: 61 n5.

[51] Donaldson (1995).

management team are what gives life to the corporation as a stable, enduring entity, which is capable of making long-term commitments.[52] One implication is that directors should do their best to ensure that key managers are well entrenched and that their wealth is tied up on a long-term basis in the companies in which they are employed. Homo economicus is not the only model of man available to agency theorists. Pepper and Gore have proposed "behavioural economic man" in their article entitled "Behavioural agency theory", arguing that agents should be modelled as "boundedly rational" (i.e., that there are neuro-physiological rate and storage limits on the ability of agents to receive, store, retrieve, and process information without error). They postulate that agents are intrinsically as well as extrinsically motivated, in such a way that in some circumstances there is a trade-off between these two types of motivation. Agents are also loss averse rather than risk averse, and hyperbolic rather than exponential time discounters, thus over-emphasising the importance of the present at the expense of the future.[53] Bosse and Phillips argue that agency theory should be developed by assuming that principals and agents exhibit "bounded self-interest"—self-interest bounded by norms of reciprocity and fairness so that agency relationships are mediated through positively and negatively reciprocal behaviours.[54] Samuel Bowles argues for "homo socialis", postulating that humans have social preferences which are affected by "altruism, reciprocity, intrinsic pleasure in helping others, aversion to inequity, ethical commitments, and other motives that induce people to help others more than is consistent with maximising their own wealth or material payoff".[55] Sigmund, Fehr, and Nowak refer to "homo emoticus", arguing, like the political scientist Jon Elster, that economic theory must take into account human emotions as well as reason.[56] Alger and Weibull propose "homo moralis", contending that a combination of selfishness and morality stands out as being evolutionarily stable.[57]

---

[52] Mayer (2013).

[53] Pepper and Gore (2015).

[54] Bosse and Philips (2016).

[55] Bowles (2016).

[56] Sigmund et al. (2001); Elster (1998). Emotions and economic theory. *Journal of Economic Literature*, 36 (1) pp. 47–74.

[57] Alger & Weibull (2013). Homo moralis – preference evolution under incomplete information and assortative matching. *Econometrica*, 81 (6) pp. 2269–2302.

There will be those who criticise this line of argument, contending that I am introducing normative considerations (of what ought to be the case) into a positive theory (of what is the case). I offer two arguments in defence. First, for all that it is often described as a positive theory,[58] agency theory has become laden with normative considerations. When Michael Jensen and Kevin Murphy calculated that the pay-performance relationship for CEOs was a $3.25 change in CEO wealth for every $1000 in shareholder wealth, their conclusion was not that the empirical evidence refuted agency theory, but that political forces operating in public markets and inside public corporations were placing constraints on pay for performance.[59] Agency theorists, therefore, recommended increasing the amount of leverage in CEO pay arrangements, an essentially normative conclusion. In Chap. 1 of *The Economics of Welfare*, A.C. Pigou argues that the distinction between positive and normative theories is often blurred in economics.[60] Kenneth Boulding maintains that the object of social science is to find out what is possible, not what merely is the case.[61]

Second, and more subtly, there is substantial evidence that some combination of "homo socialis", "homo emoticus", and "homo moralis" is much closer to the truth than "homo economicus". A number of behavioural scientists have demonstrated empirically that people are not wholly self-interested and motivated only by money—I expand on this point in Chap. 5. To model executive behaviour on the basis of a misleading set of assumptions about human behaviour risks building theory leading to flawed conclusions—the problem that Professor Tony Lawson of Cambridge University identified with the assumptions made by Milton Friedman.[62] In other words, a positive theory of agency constructed on the premise that our "model of man" should assume a combination of both knightly and knavish behaviour (i.e., that we are "boundedly self-interested" as well as "boundedly-rational") is better social science.

My objective in this book is to provide an empirically grounded and realistic theory about the relationship between company managers, shareholders, and other key corporate stakeholders which will help to guide

---

[58] Eisenhardt (1989).
[59] Jensen and Murphy (1990).
[60] Pigou, A. (1920/1932). *The Economics of Welfare*. London: Macmillan.
[61] Boulding, K. (1953). *The Organizational Revolution*. New York: Harper & Brothers.
[62] See note 16 above.

corporate action. I aim to do this by (1) starting from a realistic set of premises; (2) establishing a universally desirable objective function; and (3) proceeding logically from (1) to (2). I also aim to comply with critical realism's criteria for good scientific theorising[63]—for more on this, see Chap. 3.

CONCLUSION

To summarise the argument to this point: standard agency theory consistently confuses the positions, responsibilities and rights of shareholders, directors, and managers in public corporations.[64] Property rights within the firm are not unitary and can be easily disaggregated. Shareholders do not have a complete bundle of rights which would make them owners in any conventional sense, nor are they owners in a way which clearly distinguishes their contribution from those of certain other stakeholders who also make substantial commitments to the firm.[65] Managers and employees often dedicate large parts of their careers to a single company, developing company-specific skills that are not readily transferable—specific assets, in the words of Blair and Stout[66]—which give them a legitimate interest in the firm. Company directors are not agents of shareholders in a technical sense, but are in fact a special kind of fiduciary more akin to trustees than agents, with specific duties of care and loyalty to the firm, rather than to stockholders—the doctrine of "director primacy".[67] It is not only between shareholders, directors, executives, and other employees that chains of agency and fiduciary relationships exist. Many stockholders are themselves agents or fiduciaries, holding shares in pooled funds, and acting on behalf of retail investors, giving rise to a separate set of agency costs and risks which are too often overlooked by scholars. Finally, good corporate governance, practices, and incentive designs cannot be constructed on the basis of entirely negative assumptions about people's nature, preferences, and behaviours. Some kind of virtue ethic for agents and fiduciaries is needed to underpin the current assumptions about economic man.

---

[63] Bhaskar, R. (1975). *A Realist Theory of Science*. Leeds, UK: Leeds Books Ltd.
[64] Clark (1985).
[65] Heath (2014).
[66] Blair and Stout (1999).
[67] Bainbridge (2003).

This analysis is, in Heath's typology, an ontological theory about the fundamental nature and purpose of companies, and their managers, combining elements of deontic, governance, and corporate law stakeholder theories. It is submitted that a more complete agency model for public corporations in which institutions are properly embedded should take account of all these factors. Agency theorists have focused much attention on the use of high-powered incentives as a mechanism for overcoming agency problems within firms. In so doing, they have dramatically underestimated the role that institutions play in determining organisational behaviour. Subscribers to the standard model have spent considerable amounts of time devising highly complex incentive plans while ignoring the fiduciary responsibilities and professional ethics of directors and managers.[68]

The centrality of fiduciary relationships in this analysis illustrates the underlying paradox in standard agency theory—it relies on the ethical motives of the chairman and other non-executive directors to solve agency problems, when the conventional model of man in neoclassical economics assumes uni-dimensional self-interested utility-maximising behaviour.[69] This paradox reinforces the need to devise a more sophisticated model of economic man, one that recognises the significance of moral sentiments as well as economic impulses. It also gives force to the importance of developing normative models of executive and director behaviour that incorporate higher deontic expectations of company directors and senior executives—how can society ensure that high-performing individuals of high moral stature are recruited into these roles and given the tools they need to carry out their duties to maximum effect? These various conclusions about the standard model and the revisions that are necessary in a revised agency theory are summarised in Table 2.2.

Another element of the standard model is the assertion that firms are "legal fictions" serving as a "nexus for a set of contracting relationships among individuals".[70] As such, according to Jensen and Meckling, it makes no sense to talk about a firm's behaviour, social responsibility, or objective function. As they say: "there is in a very real sense only a multitude of complex relationships (i.e., contracts) between the legal fiction

---

[68] Heath (2014).
[69] Hodgson (2015).
[70] Jensen and Meckling (1976) pp. 310.

Table 2.2 Assumptions underpinning the standard and revised models of agency

Assumption	Standard model	Revised model
Model of man	Economic man (or homo economicus): man is rational, rent-seeking, self-interested, and there is no non-pecuniary agent motivation	Behavioural economic man: man is boundedly rational, boundedly self-interested, and intrinsically as well as extrinsically motivated
Ownership and property rights—who owns the firm?	The doctrine of shareholder primacy: shareholders are the owners of a firm	The doctrine of director primacy: shareholders hold property rights in a company's shares entitling them to dividends and certain other residual cash flows, plus representation rights
Firm's primary objective—what is a company for?	The doctrine of shareholder value maximisation: the overriding objective of a corporation's directors and managers is to maximise shareholder value	The doctrine of total firm value maximisation
Role of directors and managers	Directors and managers are agents of shareholders	Directors are a unique form of fiduciary more closely resembling trustees than agents. Managers are strictly the agents of directors
Where do agency costs arise?	Agency costs arise because of the different interests of shareholders and managers	Agency costs arise because of the different interests of retail investors and funds, shareholders and directors, directors and managers, and managers and other employees
Mechanism for allocating residual profits	Shareholders are entitled to residual profits, which they receive by way of dividends and capital gains	Residual profits should be shared by all stakeholders who make specific personal investments in the firm

(the firm) and the owners of labor, material and capital inputs and the consumers of output".[71] This a strange way of putting it, if you think about it, as a *fiction* surely cannot enter into a contract; what I think they really mean is that the contracts are all between natural persons (i.e., directors, executives, employees, shareholders, suppliers, customers, etc.,) which we conceptualise as "a firm" much as the philosopher Gilbert Ryle conceptualises the mind as a "ghost" in the "machine" of the body;[72] in other words the firm, like the mind according to Ryle, does not really exist. Jensen and Meckling advance no arguments in support of this claim, thus ignoring nearly 200 years of philosophical argument about the nature of companies. That assertion requires further investigation; therefore, I turn next to the question of "what a public corporation really is" in Chap. 3.

Further Reading
The key readings on the standard model can be found in: Kroszner, R., & Putterman, L. (2009) *The Economic Nature of the Firm – A Reader. Third Edition.* Cambridge University Press. A valuable collection of accessible essays on agency theory is: J. Pratt & R. Zeckhauser (Eds.), *Principals and Agents: the Structure of Business.* Harvard Business School Press. Margaret Blair's and Lynn Stout's work on the "team production theory of company law" is summarised in: Stout, L. (2012). *The Shareholder Value Myth: How Putting Shareholders First Harms Investors, Corporations, & the Public.* Berrett-Koechler Publishers, Inc.

REFERENCES

Aguinis, H., Gomez-Mejia, L., Martin, G., & Joo, H. (2018). CEO Pay is Indeed Decoupled from CEO Performance: Charting a Path for the Future. *Management Research, 16*(1), 117–136.

Alchian, A., & Demsetz, H. (1972). Production, Information Costs and Economic Organization. *American Economic Review, 62*(5), 777–795.

Alger, I., & Weibull, J. (2013). Homo Moralis: Preference Evolution Under Incomplete Information and Assortative Matching. *Econometrica, 81*(6), 2269–2302.

---

[71] Jensen and Meckling (1976) pp. 311.
[72] Ryle, G. (1949/2000). *The Concept of Mind.* London: Penguin Books.

Bainbridge, S. (2003). Director Primacy: The Means and Ends of Corporate Governance. *Northwestern University Law Review, 97*(2), 547–606.

Baker, G., Jensen, M., & Murphy, K. (1988). Compensation and Incentives: Practice vs. Theory. *Journal of Finance, 43*(3), 593–616.

Bebchuk, L., & Fried, J. (2004). *Pay Without Performance – The Unfulfilled Promise of Executive Compensation.* Cambridge, MA: Harvard University Press.

Besley, T., & Ghatak, M. (2005). Competition and Incentives with Motivated Agents. *American Economic Review, 95*(3), 616–636.

Bhaskar, R. (1975). *A Realist Theory of Science.* Leeds: Leeds Books.

Blair, M. (1995). *Ownership and Control: Rethinking Corporate Governance for the Twenty-first Century.* Washington, DC: The Brookings Institution.

Blair, M. (1996). *Wealth Creation and Wealth Sharing.* Washington, DC: The Brookings Institution.

Blair, M., & Stout, L. (1999). A Team Production Theory of Corporate Law. *Virginia Law Review, 85*(2), 247–328.

Bosse, D., & Philips, R. (2016). Agency Theory and Bounded Self-interest. *Academy of Management Review, 41*(2), 276–297.

Boulding, K. (1953). *The Organizational Revolution.* New York: Harper & Brothers.

Bowles, S. (2016). *The Moral Economy – Why Good Incentives are no Substitute for Good Citizens.* New Haven: Yale University Press.

Clark, R. (1985). Agency Costs Versus Fiduciary Duties. In J. Pratt & R. Zeckhauser (Eds.), *Principals and Agents: The Structure of Business.* Boston: Harvard Business School Press.

Davis, J., Schoorman, F., & Donaldson, L. (1997). Toward a Stewardship Theory of Management. *Academy of Management Review, 22*(1), 20–47.

Donaldson, L. (1995). *American Anti-Management Theories of Organization.* Cambridge: Cambridge University Press.

Donaldson, L., & Davis, J. (1991). Stewardship Theory or Agency Theory: CEO Governance and Shareholder Returns. *Australian Journal of Management, 16*(1), 49–64.

Easterbrook, F., & Fischel, D. (1948/1991). *The Economic Structure of Corporate Law.* Cambridge, MA: Harvard University Press.

Edmans, A., & Gabaix, X. (2016). Executive Compensation: A Modern Primer. *Journal of Economic Literature, 54*(4), 1232–1287.

Eisenhardt, K. M. (1989). Agency Theory: An Assessment and Review. *Academy of Management Review, 14*(1), 57–74.

Elster, J. (1998). Emotions and Economic Theory. *Journal of Economic Literature, 36*(1), 47–74.

Freeman, R. (1984/2010). *Strategic Management: A Stakeholder Approach.* Cambridge, UK: Cambridge University Press.

Friedman, M. (1953/2008). The Methodology of Positive Economics. In D. Hausman (Ed.), *The Philosophy of Economics: An Anthology* (3rd ed., pp. 145–178). Cambridge: Cambridge University Press.

Friedman, M. (1970, September 13). The Social Responsibility of Business Is to Increase Its Profits. *The New York Times Magazine.*

Frydman, C., & Saks, R. (2010). Executive Compensation: A New View From a Long-term Perspective, 1936–2005. *The Review of Financial Studies, 23*(5), 2099–2138.

Gabaix, X., & Landier, A. (2008). Why Has Executive Pay Increased so Much? *Quarterly Journal of Economics, 123*(1), 49–100.

Ghoshal, S. (2005). Bad Management Theories Are Destroying Good Management Practices. *Academy of Management – Learning & Education, 4*(1), 75–91.

Granovetter, M. (1985). Economic Action and Social Structure: The Problem of Embeddedness. *American Journal of Sociology, 91*(3), 481–510.

Grossman, S., & Hart, O. (1983). An Analysis of the Principal-agent Problem. *Econometrica, 51*(1), 7–45.

Harris, J., Johnson, S., & Souder, D. (2013). Model-theoretic Knowledge: The Case of Agency Theory and Incentive Alignment. *Academy of Management Review, 38*(3), 442–454.

Heath, J. (2014). *Morality, Competition, and the Firm.* New York: Oxford University Press.

Hodgson, G. (2015). *Conceptualizing Capitalism: Institutions, Evolution, Future.* Chicago: University of Chicago Press.

Honore, A. (1961). Ownership. In A. Guest (Ed.), *Oxford Essays in Jurisprudence* (pp. 107–147). Oxford: Oxford University Press.

Hume, D. (1804). *Of the Independency of Parliament Essays, Moral, Political, and Literary.* Edinburgh: Bell & Bradfute.

Jensen, M. (2001). Value Maximization, Stakeholder Theory, and the Corporate Objective Function. *Journal of Applied Corporate Finance, 14*(3), 8–22.

Jensen, M., & Meckling, W. (1976). Theory of the Firm: Managerial Behavior, Agency Costs and Ownership Structure. *Journal of Financial Economics, 3*(4), 305–360.

Jensen, M., & Murphy, K. (1990). Performance Pay and Top-Management Incentives. *Journal of Political Economy, 98*(2), 225–264.

Kay, J. (2012). The Kay Review of UK Equity Markets and Long-term Decision Making. *Final Report.* Retrieved from London: https://www.gov.uk/government/uploads/system/uploads/attachment_data/file/253454/bis-12-917-kay-review-of-equity-markets-final-report.pdf

Keay, A. (2013). *The Enlightened Shareholder Value Principle and Corporate Governance.* Abingdon: Routledge.

Lawson, T. (2015). *Essays on the Nature and State of Modern Economics.* Abingdon: Routledge.

Le Grand, J. (2003). *Motivation, Agency and Public Policy.* Oxford: Oxford University Press.

Mackenzie, D. (2007). Is Economics Performative? In D. Mackenzie, F. Muniesa, & S. Lucia (Eds.), *Do Economists Make Markets?* Princeton: Princeton University Press.

Mayer, C. (2013). *Firm Commitment: Why the Corporation Is Failing Us and How to Restore Trust in It.* Oxford: Oxford University Press.

Milgrom, P., & Roberts, J. (1992). *Economics, Organisation and Management* (2nd ed.). Englewood Cliffs: Prentice-Hall Inc.

Nowak, M., & Highfield, R. (2011). *Super Cooperators – Evolution, Altrusim and Human Behaviour.* Edinburgh: Canongate.

Nowak, M., Page, K., & Sigmund, K. (2000). Fairness Versus Reason in the Ultimatum Game. *Science, 289*(5485), 1773–1775.

Orts, E. (2013). *Business Persons: A Legal Theory of the Firm.* Oxford: Oxford University Press.

Pepper, A., & Gore, J. (2015). Behavioral Agency Theory: New Foundations for Theorizing About Executive Compensation. *Journal of Management, 41*(4), 1045–1068.

Pigou, A. (1920/1932). *The Economics of Welfare.* London: Macmillan.

Ross, S. (1973). The Economic Theory of Agency: The Principal's Problem. *American Economic Review, 63*(2), 134–139.

Ryle, G. (1949/2000). *The Concept of Mind.* London: Penguin Books.

Sigmund, K., Fehr, E., & Nowak, M. (2001). The Economics of Fair Play. *Scientific American, 286*(1), 82–87.

Spence, M., & Zeckhauser, R. (1971). Insurance, Information and Individual Action. *American Economic Review, 61*(2), 380–387.

Stout, L. (2012). *The Shareholder Value Myth: How Putting Shareholders First Harms Investors, Corporations, and the Public.* San Francisco: Berrett-Koechler Publishers.

Tosi, H., Werner, S., Katz, J., & Gomez-Mejia, L. (2000). How Much Does Performance Matter? A Meta-Analysis of CEO Pay Studies. *Journal of Management, 26*(2), 301–339.

van Essen, M., Otten, J., & Carberry, E. (2015). Assessing Managerial Power Theory: A Meta-Analytic Approach to Understanding the Determinants of CEO Compensation. *Journal of Management, 26*(2), 164–202.

Williamson, O. (1975). *Markets and Hierarchies.* New York: The Free Press.

Williamson, O. (2011). Corporate Governance: A Contractual and Organizational Perspective. In L. Sacconi, M. Blair, R. Freeman, & A. Vercelli (Eds.), *Corporate Social Responsibility and Corporate Governance.* Basingstoke: Palgrave Macmillan.

CHAPTER 3

What a Public Corporation Really Is

Abstract This chapter addresses one of the assumptions of standard agency theory—that the corporation is a legal fiction. It sets out a series of arguments as to why this assumption is incorrect, explains why public corporations should be regarded in ontological terms as real entities, and spells out why this matters—the consequences for agency theory as a whole and the particular consequences for shareholders and directors.

Keywords Critical realism • Social ontology • Public corporation

INTRODUCTION

In 1916 the sociologist Harold Laski wrote:

> The state knows certain persons who are not men. What is the nature of their personality? Are they merely fictitious abstractions, collective names that hide from us the mass of individuals beneath? Is the name that gives them unity no more than a convenience, a means of substituting one action in the courts where, otherwise, there might be actions innumerable? Or is that personality real? Is Professor Dicey right when he urges that "whenever men act in concert for a common purpose, they tend to create a body which, from no fiction of law but from the very nature of things, differs from the

individuals of whom it is constituted"?[1] Does our symbolism, in fact, point
to some reality at the bottom of appearances? If we assume that reality, what
consequences will flow therefrom?

Laski, H. (1916). The Personality of Associations. *Harvard Law Review,*
29(4), 404–426

In the previous chapter I explained that, in the economists' standard
model, a firm is regarded as a "legal fiction" and a "nexus for a set of contracting relationships".[2] This is consistent with "methodological individualism", an approach adopted by most economists and many other social
scientists, which postulates that causal accounts of social phenomenon are
the result of the actions of individual agents. According to the "nexus of
contracts" view, the fundamental components of the firm are individual
shareholders, directors, managers and employees, and the relationships
between them that are governed by contract. There is no essential difference in this respect, so the "nexus of contracts" proponents say, between
a closely held private company and a public corporation.

This approach has been criticised by scholars from a number of different disciplines: by sociologists, who argue that the standard model is
under-socialised, and who focus on the omission of important sociological
concepts like power, politics, and culture;[3] by philosophers, who reflect on
issues of identity and moral personality;[4] and by legal scholars who place
the firm in a historical context, examining how the legal conception of a
firm (be it a private company, joint-stock company, public corporation,
charter company, or whatever) has evolved over time and has been affected
by case law precedents, company law statutes, and so on.[5]

Management scholars have also criticised the "legal fiction" and "nexus
of contracts" view of the firm, arguing that it is one of a number of "bad
management theories" which potentially undermine "good management
practices"[6] and that the negative view of human nature which is implicit in

[1] Albert Venn Dicey (1835–1922), Professor of English Law at Oxford, was a leading
constitutional scholar of his day.

[2] Jensen, M., & Meckling, W. (1976). Theory of the firm: Managerial behavior, agency
costs and ownership structure. *Journal of Financial Economics,* 3(4) pp. 305–360.

[3] E.g., Perrow, C. (1986). *Complex Organizations.* New York: Random House.

[4] E.g., Scruton, R. (1990). Gierke and the corporate person *The Philosopher on Dover
Beach.* New York: St Martin's Press.

[5] E.g., Bratton, W. (1989). The new economic theory of the firm: critical perspectives from
history. *Stanford Law Review, 41*(6), 1471–1527.

[6] Ghoshal, S. (2005). Bad management theories are destroying good management practices. *Academy of Management – Learning & Education, 4*(1), 75–91.

the standard model fails to recognise positive traits which are also demonstrated by managers in practice—it therefore runs the risk of becoming a self-fulfilling prophecy.[7] As well as being a practice, management is a field of study, not a discipline like economics or law, which have their own unique methods of enquiry. Just as philosophers are, in John Locke's terms, "under-labourers" for the natural and social scientists,[8] so there is a sense in which management scholars must look to other social scientists to explicate the fundamental concepts with which they work.[9] Accordingly, I now turn to philosophy, law, economics, and sociology to find an answer to the question: "what a company really is?"

CRITICAL REALISM

Perhaps the most interesting critique of the "nexus of contracts" and "legal fiction" view of the firm has come from within the economics profession itself, specifically from a group of economists associated with Professor Tony Lawson of Cambridge University, and the Cambridge Social Ontology Group. This research group was formed with the specific aim of systematically studying the nature and basic structure of social reality, a branch of philosophical investigation known as "social ontology". The group is closely associated with an approach to social ontology known as "critical realism", which was pioneered by the philosopher Roy Bhaskar and combines a general philosophy of science (transcendental realism) with a philosophy of social science (critical naturalism) to describe the interface between the physical and social worlds.[10] Critical realism departs from methodological individualism in that it postulates that social objects really exist as entities which are separate from the individuals with whom they are associated. Reality looks something like an onion with many different layers (see Fig. 3.1).

[7] Donaldson, L. (1995). *American Anti-management Theories of Organization*. Cambridge: Cambridge University Press.

[8] Locke, J. (1690 | 1960). *An Essay Concerning Human Understanding*. Glasgow: Collins.

[9] Willman, P. (2014). *Understanding Management – the Social Science Foundations*. Oxford: Oxford University Press.

[10] Bhaskar, R. (1975). *A Realist Theory of Science*. Leeds, UK: Leeds Books Ltd., and Bhaskar, R. (1979). *The Possibility of Naturalism – a Philosophical Critique of the Contemporary Human Sciences*, Brighton, UK: Harvester Press.

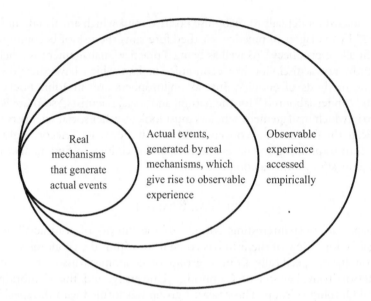

Fig. 3.1 The nature of reality in critical realism

The outer layer represents empirically observable experiences. Behind experiences lie actual events that give rise to the observed phenomena. Behind events are the real mechanisms that have generated the events. Scientists use many different kinds of theories, constructs, algorithms, models, and metaphors in trying to explain these events. We try to strip away the layers of reality, but often fail, and are constrained by problems of perception. This is why we make use of models, metaphors, and other mental constructs to help us order chaotic sense data and explain actual events. We are forced to do this because the generative mechanisms are transcendent, beyond direct human experience; however, they are not necessarily beyond human knowledge, because philosophical argument and the powers of imagination may help us to transcend the bounds of perception.

Different beliefs about the discovery process in the modern social sciences sometimes give rise to academic tournaments, pitting on one side those who believe in an essentially inductive approach, beginning with empirical data and, through a process of searching for correlations and other patterns, constructing models, and theories, against those on the other side who believe in an essentially deductive approach, beginning

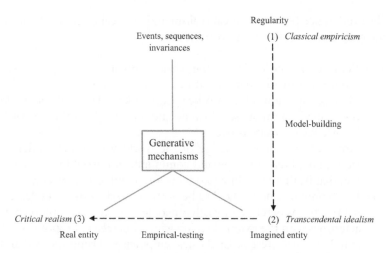

Fig. 3.2   The process of scientific discovery in critical realism (after Bhaskar, 1975)

with a postulated model or theory which they then test against the data. An example of the former might be the psychologist who, starting with a bunch of experimental data, builds a set of explanatory constructs. An example of the latter might be the economist who begins her research with a mathematical model, which she then tests against empirical data. Philosophically speaking, the first approach is closest to classical empiricism, starting as it does with sense data from which theory is developed. The second approach is closest to idealism, starting with a priori postulates before proceeding to the data.

For the critical realist, the process of scientific discovery proceeds in a different way. It begins, in Bhaskar's words, with: "a kind of dialectic in which a regularity is identified"...[and]... "a plausible explanation for it is invented".[11] This two-step process is initially similar to the approach taken by the classical empiricist, whereas the idealist works in the opposite direction, proceeding from theory to facts. But the critical realist takes the third step, proceeding from initial observations to possible explanations (theories), before testing: "the reality of the entities and processes postulated in the explanation" (see Fig. 3.2).

---

[11] Bhaskar (1975) p. 145.

The difference between critical realism and classical empiricism can be made clear if we consider four propositions:

(1) Because my knowledge of the world is obtained via my senses, I cannot be sure of what constitutes external reality.

(2) Because my knowledge of other people's knowledge of the world is obtained via my senses, I cannot be sure of what other people's knowledge of reality is like.

(3) However, if your knowledge of the world was entirely different from mine, then language and communication would be impossible, that is, there would be no basis for "common knowledge".

(4) For "common knowledge" to be possible, what constitutes "external reality" as perceived by other people must have common characteristics with the external reality which I perceive. Propositions 1 and 2 are characteristic of classical empiricism. Propositions 3 and 4 go further and constitute critical realism.

The third step in Bhaskar's method is in many ways the most difficult to understand. He describes it in the following way:

> For transcendental realism that some real things and generative mechanisms must exist can be established by philosophical argument (their existence, and transfactual activity, is a condition of the possibility of science). But it is contingent and the job of substantive science to discover which hypothetical or imagined mechanism are not imaginary but real; or, to put it the other way round, to discover what the real mechanisms are, i.e., to produce and adequate account of them.[12]

This third step involves an act of philosophical projection, because the underlying mechanisms may not (and arguably in most circumstances cannot) be observable. The task of the scientist, using experiments and controls wherever possible, is to establish which of the hypothetical mechanisms are likely to be real rather than hypothetical, recognising that the answer will inevitably be contingent, probabilistic, and something to be viewed with an appropriate degree of scepticism. Absolute certainty can never be

[12] Bhaskar (1975) p. 146.

achieved; the true scientist must be prepared to switch theories in the face of new data or superior argument.

Heady stuff, but what does all this mean when applied to social constructs such as "firm", "company", or "public corporation"? Here I must first insert a note on terminology. Economists favour the use of the word "firm", referring generally to a business enterprise; this might, more specifically, refer to a sole trader, business partnership, or, in UK parlance, a company. A company might be a private company (e.g., a family-owned company or a subsidiary company which is a member of a larger group) or a public company (whose shares will often, though not necessarily, be listed on a stock exchange and publically traded). Companies may in turn be limited (normally by shares, but occasionally by guarantee) or unlimited (depending on whether the liability of the members is capped or not). While most companies in the UK are incorporated under the Companies Acts, some, especially for historical reasons, are statutory companies established by special Act of Parliament or chartered companies established under special powers conferred upon the Crown. Companies may be "for profit" or "not for profit" depending upon whether commercial or other objectives are their primary stated aims, and, for accounting purposes, may be classified as "large", "medium", or "micro".

In the US the term "company" might also refer to a partnership or some other form of collective ownership, and even sometimes to a sole proprietorship. Incorporated companies are generally referred to as "corporations". "Limited liability companies" (LLCs) are, in effect, sole proprietorships (they are taxed as such) but provide limitations on the proprietor's liability. Each American state and territory has its own basic corporate legal code, and corporations are normally incorporated under state law, while federal law sets the standards for trading in shares and corporate governance. As in the UK, there are other varieties of corporate form.

I make two points about these various fine distinctions before moving on. Firstly, the neoclassical economists' general "theory of the firm" would appear to treat certain entities in generic terms when they may, in ontological terms, be quite different. Secondly, the ontological status of the various types of the corporate form is not necessarily the same; to use a biological metaphor, "companies" and "corporations" may refer to equivalent orders or families but not necessarily to the same genus or species. Careful examination of both the law and the facts is required.

For present purposes, given the particular focus of this chapter, I can now simplify matters by distinguishing between (1) large, for profit, public limited companies, or groups of companies, with shares listed on a stock exchange, which I refer to as "public corporations"; (2) private limited companies whose shares are closely held ("private companies"); and (3) all other types of "company" or "firm".[13] The focus of this chapter, and indeed the whole book, is primarily on the first of these categories, public corporations.

The Social Ontology of Public Corporations

A good place to start this critical examination of what is meant by "public corporation" are three questions posed by legal scholar William Bratton.[14] (1) Is a corporation a "reification", that is to say, a construction of the minds of persons connected with the firm, or a real thing, having a separate existence from such persons?[15] (2) What is the distinction between a corporation, on the one hand, and both the individuals associated with it and the transactions, which go on in and around it, on the other hand? In other words, is a corporation a single entity or an aggregate of separate entities (as it would be if it was a nexus of contracts)? (3) To what extent does a corporation derive its authority from the state? Another way of putting this is to draw a distinction between the public and the private—is a corporation a public body in which the state and persons generally have an interest, or is it a separate, independent body whose activities are governed by private law?

Bratton goes on to address these questions using nineteenth- and twentieth-century American legal history as his field of enquiry, and I will come to this in a moment. First, I take as a second point of departure a definition of the firm provided by the economist Tony Lawson from a

---

[13] Even these categorisations are not ideal. Some very large corporations are privately held, e.g., Bechtel Corporation and Koch Industries in the US, and some very large enterprises are actually partnerships e.g., PwC.

[14] Bratton, W. (1989). The new economic theory of the firm: critical perspectives from history. *Stanford Law Review, 41*(6), 1471–1527.

[15] Bratton even poses, as a subsidiary question, whether it would make sense to describe a corporation as a spiritual being.

critical realist perspective.[16] Lawson defines a firm in a generic sense (and I paraphrase) as comprising *people* and *artefacts* and an organising *relationship structure* between them. The organising structure involves social positions (which he calls *positional identities*), rights and obligations. These positions and identities are in part determined by *legal actions*, such as the process of incorporation, and *legal institutions*, including company law, employment law, and tax law in general, as well as the corporation's memorandum and articles of association (UK) or constitution and by-laws (US) in particular.[17] The organising structure may be *emergent*, which is to say, subject to evolutionary processes.

Lawson argues that all social entities that include natural persons (i.e., individual human beings) as elemental components are examples of *communities*, which he regards as being a fundamental type of social entity. Within a community people may have offices or roles, where the office or role is defined separately from the office-holders or role-holders. If someone is positioned in a community as, for example, a parish priest,[18] then the role is not determined by the personality or capacities of the individual who holds the office at any particular time, but rather that the individual becomes the bearer of a set of positional powers, rights and obligations. So in a public corporation, an individual person may be positioned as the chairman, with certain powers, rights, and obligations, as the chief executive officer, with other powers, rights and obligations, and so on. These powers, rights and obligations are substantially determined by the applicable law, so an analysis of such legal attributes is critical to understanding the nature of the corporation. The structures are inevitably contingent and the range of possibilities will vary according to the applicable laws in different jurisdictions as well as over time. So, for example, a firm that

---

[16] Lawson (2015a). A conception of social ontology. In S. Pratten (Ed.), *Social Ontology and Modern Economics* (pp. 19–52). London and New York: Routledge, and Lawson (2015b). The nature of the firm and peculiarities of the corporation. *Cambridge Journal of Economics, 39*(1), 1–32.

[17] A corporation's memorandum of association sets out the company name, location (registered office and country of domicile), founding members, and main purpose. The articles of association set out the kind of business to be undertaken, the responsibilities of the directors, and the means by which the shareholders exercise control over director. Similar matters are dealt with in a US corporation's constitution and by-laws.

[18] The English parish priest is one of the examples of a "corporation sole" which the legal historian F.W, Maitland (1850–1906) examines in Chapter 1 of his book *State, Trust and Corporation* (1911 | 2003).

starts life as an unincorporated business with a sole proprietor may subsequently become a limited company with a number of shareholders separate from the founder, and might eventually become a public limited company (UK) or public corporation (US) with multiple shareholders. In this process of evolution, the underlying reality of the firm may change.[19]

Lawson also argues more generally that social reality is an open-ended system, not a closed system—that it is a process, in motion, whose parts are constituted by their changing relations with each other, that a social category must pick out a definite referent, that categories should be consistent with historical usage, and that the conception of a social category that is to be defended must have some theoretical or practical utility.[20]

DEVELOPMENT OF COMPANY LAW IN THE UK AND THE US AFTER 1800

Having done the groundwork, I now return to Bratton's historical analysis of the emergence of the law relating to public corporations in the US in the nineteenth and early twentieth centuries, which I combine with an analysis of the development of UK company law, drawing on the work of the economic historian, Phillip Cottrell.[21]

Until the early part of the nineteenth century, business firms in the US and UK could, legally speaking, be organised in one of three separate ways: as a partnership (or "co-partnery", in the words of Adam Smith), as an unincorporated company, or as a corporation (i.e., an incorporated

[19] One mistake which I believe Lawson makes is to write about the social ontology of companies in general terms, without recognising that a large public corporation appears prima facie to be a very different thing, ontologically speaking, from a small private company. He also privileges tax law in determining the nature of the company, which seems peculiar given the enormous body of other law that applies to companies, and the great range in tax regimes which can be found across different jurisdictions. It is for these reasons that I have taken Lawson's definition of firms and corporations as a starting point and performed a historical and philosophical analysis of my own.

[20] Lawson (2015a).

[21] Cottrell (1979) *Industrial Finance 1830–1914. The Finance and Organization of English Manufacturing Industry*. Methuen. As well as Cottrell and Bratton, I am also heavily indebted in this historical section to Harris, R. (2000) *Industrializing English law. Entrepreneurship and business organization, 1720–1844*. Cambridge, UK: Cambridge University Press, and Micklethwait, J., & Wooldridge, A. (2003). *The Company. A Short History of a Revolutionary Idea*. London: Weidenfeld & Nicolson.

company). Partnerships were organised under contract law, and unincorporated companies were organised by vesting assets under the supervision of trustees by deeds of settlement.[22] Corporations were subject to the state, being formed by Act of Parliament, or by being given a legal charter by Parliament or the Crown. The earliest examples of large corporations were the great chartered trading companies, for example, the Muscovy Company (a joint-stock company founded in 1551 and given a Royal Charter by Mary I of England in 1555), the East India Company (founded in 1600), and the Hudson's Bay Company (founded 1670). Provision of a charter was often accompanied by the granting of monopoly trading rights over a particular geographical region (e.g., Russia in the case of the Muscovy company). The East India Company, along with the Bank of England (founded 1694) and South Sea Company (founded 1711), became known as the "Moneyed Companies" because of the key role which they came to play in helping the UK government to manage its debt.[23] The last of these was involved in an infamous early state-wide financial crisis, the "South Sea Bubble" of 1720, when unwarranted speculation in the company's shares, egged on by parliamentarians, led to a major financial crisis. This has been described by Richard Dale as "The first [financial] crash".[24]

Between 1825 and 1862 a series of Acts of Parliament revolutionised English company law and paved the way for the development of public corporations. In 1825 the Bubble Act, which dated back to 1720 and which had placed severe restrictions on the establishment of joint-stock companies, was finally repealed. The Companies Act of 1844 allowed the formation of joint-stock companies provided that they were regulated by the state, although there was no provision for limited liability which at the time was widely associated with monopoly power and speculative activity and hence was regarded with suspicion by businessmen and the judiciary alike. In 1855 and 1856 the law was further liberalised, and limited liability became more freely available, the only condition being the filing of a

---

[22] Hence the common description, especially in the US, of unincorporated companies as "trusts". Trust were also used by the likes of John D. Rockefeller and J.P. Morgan to consolidate power over groups of otherwise separate corporations.

[23] Harris (2000).

[24] Dale, R. (2004). *The First Crash – Lessons from the South Sea Bubble*. Princeton, NJ: Princeton University Press.

memorandum of association with the Registrar of Joint-Stock Companies.[25] The relevant law was consolidated in the Companies Act 1862, by which time English company law had become "the most permissive in Europe".[26] A series of subsequent Companies Acts up to the present day have essentially retained the same fundamental features.

In the early part of the eighteenth century, developments in American company law closely paralleled those in English law, with which, for historical reasons, American law was of course closely entwined. The law regarded individuals rather than collectives as the fundamental units of legal analysis, so that contracting and market organisation was predominant. Corporate legal doctrine, inherited from Great Britain and based on "concession theory" (which presupposes that a juristic person is merely a concession or creation of the state), held that the corporate form should be made available principally by the sovereign grant of charter. New grants were to be permitted only in certain circumstances, for example, state franchises, natural monopolies, public utilities, transport, and so on. Unincorporated companies were regarded as "legal fictions" and "artificial entities". Contractarian principles also applied—firms were regarded as aggregate collections of individuals; incorporation merely cemented the fiction. For example, in the judgment in the Dartmouth College case of 1819 it was held that: "a corporation is an artificial being, invisible, intangible, and existing only in contemplation of law".[27]

Things changed substantially after 1850 with the development of manufacturing industry and the factory economy. American states enacted general corporate laws to ensure that there would be ready access to the corporate business format. These laws dealt with a company's purpose and objectives, capital structure, dividends, directors' duties, and so on. Corporations had governance structures that were generally relatively simple, uniform hierarchies, thus known as "U-form" companies, although with the growth of the railroad networks the first of the great management hierarchies appeared, these being according to Bratton: "the most striking departure from the classical economic model" as it had been espoused by

---

[25] See the Limited Liability Act 1855 and Joint Stock Companies Act 1856.
[26] Cottrell (1979) p. 52.
[27] Dartmouth College v. Woodward, 17 US (4 Wheat.) 518, 636 (1819), cited by Bratton, 1989.

Adam Smith.[28] However, unlike later managerial corporations, which had large numbers of outside shareholders with small stock-holdings, the railroads tended to have a small number of large stockholders with sizable blocks of shares. In terms of corporate doctrine, the notions of "legal fiction" and "artificial entity" were called into question and gradually replaced. Ready access to the corporate form, without the need for charters or acts of parliament, meant that corporations no longer appeared to be so closely associated with the state, and corporate legal entities began to acquire some kind of social reality.

Managerial corporations, organised around hierarchies of executives, first appeared around 1890. Managers took on multiple tasks of production, marketing, sales, and distribution. Increasing capital requirements in asset-intensive industries led to more widely dispersed shareholders, accessed via the capital markets. The States of New Jersey and Delaware, in particular, set out to establish standardised corporate legal structures to support the growth of the new industries. In terms of legal theory, "corporate realism" (the idea that a corporation has a real existence, independent of its directors and members, and is not based on any kind of fiction) came to replace contractarianism (the notion that companies were merely aggregates of individuals bound by contract).

These developments continued in both the US and Great Britain after the First World War with the establishment of multidivisional, or "M-form" companies, pioneered in particular by Alfred Sloan at General Motors. Top management, responsible for strategy, monitoring, and investor relations, became separated from divisional management, which was responsible for operational performance. As well as General Motors, which had grown exponentially under Sloan's leadership following the acquisitions of Cadillac, Buick, Oldsmobile, Oakland, and so on, in Europe these included Unilever, formed in 1929 through the merger of the operations of the Dutch company Margarine Unie and the British soapmaker Lever Brothers.[29] The size of corporations increased because the new governance structures allowed businesses to be added in the form of new divisions without causing significant risk of loss of control. Between 1945 and 1980 General Motors' sales revenues grew from $ 3.1 billion to $57.7

---

[28] Bratton (1989) p. 1486.
[29] Sloan, A. (1964). *My Years With General Motors.* Garden City, New York: Doubleday & Company, Inc.; Wilson, C. (1954). *The History of Unilever.* London: Cassell.

billion;[30] Unilever's sales grew from $150 million to $1 billion during the same period.[31]

The M-form corporation became widespread in the US and Great Britain, especially after the Second World War, reaching a peak in the 1960s–1970s with the predominance of large industrial conglomerates. Imperial Chemical Industries (ICI), a sprawling conglomerate of chemical, pharmaceutical, and agricultural manufacturing divisions, was the largest company in Britain for much of its history from 1926 to the early 1990s when it fought off a takeover from Hanson Trust plc., which led to the divestment of its agricultural operations (sold to Norsk Hydro in 1991), nylon business (sold to DuPont in 1992) and the demerger of its pharmaceuticals business (to Zeneca in 1993), leaving ICI to focus on its speciality chemicals business. At its peak in 1975, it employed 201,000 people across the world. Two years after the demerger this number had fallen to below 65,000. The remaining part of ICI was sold to AkzoNobel in 2008.

DEVELOPMENTS IN CONTINENTAL EUROPEAN JURISPRUDENCE AFTER 1800

The development of Anglo-American company law has parallels in developments in the legal philosophy and social theory of law,[32] which took place in Germany and France in the latter part of the nineteenth and early part of the twentieth centuries. By 1900 two separate paradigms had come to dominate legal philosophers' thinking about the corporation. On one side the anti-realists, associated in particular with the German jurist and legal historian Friedrich Carl von Savigny (1779–1861), argued that corporations were fictitious, artificial creations which existed only in the eyes of the law. Savigny belonged to the German historical school of jurists who believed that the law was a product of organic growth, and who

---

[30] Sloan (1964) p. 214 and *New York Times*, February 3, 1981.

[31] Jones, G. (2002). Unilever in the United States, 1945–1980. *Business History Review*, 76 (3) p. 445.

[32] The study of legal philosophy and the social theory of law is more appropriately known as "jurisprudence" and its scholars are known as "jurists". Historical jurisprudence was particularly strong in Germany and France in the nineteenth century given the codification of law based on Roman law and Code Napoléon, in contrast to the common law tradition of "judge made law" in England and the US.

opposed the ideas of French eighteenth-century jurists and the English utilitarian Jeremy Bentham, who believed that law could be imposed irrespective of the country's history and stage of development. Savigny's view of corporations as personae fictae was derived from the Roman law of associations. While Roman law recognised the *universitas*, a separate legal entity which could hold property and was distinct from its members, as well as the *societas*, which was based on contractual relationships between members and could only hold assets collectively, subject to the contracts between its members, there was no recognition of separate corporate personality. In much the same vein, Samuel von Pufendorf (1632–1694) had previously described associations as "persona moralis composita" (composite moral persons) and Wilhelm von Humboldt (1767–1835) had said such composite persons "should be regarded as nothing more than the union of members at a given time".[33] This way of thinking became known as fiction theory and was frequently combined with concession theory, the notion that the legal fiction of the corporation only acquires status by reason of some action of the state or its sovereign. The alternative to concession theory, known by some as aggregation theory and illustrated by the quotation from Humboldt, was that corporations were aggregations of their members and that legal recognition was merely incidental.

On the other side were the corporate realists, most closely associated with the Otto von Gierke (1841–1921), and in particular with the ideas contained in his four-volume magnum opus *Das deutsche Genossenschaftsrecht* (The German Law of Associations). Gierke, like Savigny, a German legal scholar and historian, also argued that law developed organically from its historical origins, but he emphasised a distinctively German historical jurisprudence, as opposed to any Roman law foundations. Gierke claimed that the correct path of German law, as it related to community associations, could be traced back to medieval times, and he argued, based on his analysis of these historical community associations, that corporations should be regarded as real persons, on a par with "natural persons" such as individuals.

Three significant propositions are evident in Gierke's work.[34] First, he proposes that civil society, as an organising social mechanism with its own institution-building powers, exists separately from the state. Secondly, he

---

[33] Cited in Scruton (1990) p. 63.
[34] Scruton (1990) p. 59.

postulates that the law, which provides a source of authority in society, evolves from informal social institutions without the necessity of any reference to a sovereign legislator. Thirdly, he argues that, if we are to provide a positive general theory of nations and society, then it is necessary to postulate the existence of autonomous associations that are real entities and which have personalities, wills, rights and obligations. Frederick Hallis in his book, *Corporate Personality, A Study in Jurisprudence*, describes the last of these, Gierke's major metaphysical claim, in the following terms: "Roman jurisprudence had allowed the *societas* and the *universitas* to grow apart, until it becomes impossible to comprehend them as phenomena of a single social world except by means of crude fiction. It could not sketch society as a legal organisation without producing a caricature of the facts of social life. Between the omnipotent state and the single individual it could see nothing but a collection of juristic constructions".[35] Gierke's argument targets methodological individualism with an argumentum ad absurdum: because it is impossible to believe a social theory which admits only individuals and the state, and does not allow for intermediate social objects which are real entities, then we must conclude that such real entities (i.e., and not just "crude fictions") do indeed exist.

Gierke argues that alongside the general accepted sociological categories of Gemeinschaft (commonly translated as "community") and Gesellschaft (sometimes translated as "society", but also carrying connotations of "partnership" as in a business partnership) we need to postulate a third category, which he calls Genossenschaft (sometimes translated as "fellowship" or "cooperation").[36] Ferdinand Tonnies (1855–1936), the author of the original Gemeinschaft-Gesellschaft dichotomy, distinguished between social groups based on personal social interaction (Gemeinschaft), and groups constructed from rules, formal roles, and indirect transactions (Gesellschaft). The sociologist Max Weber (1864–1920), in his magnum opus *Wirtschaft und Gesellschaft* (Economy and Society), regards Gemeinschaft as being rooted in subjective feeling, whereas Gesellschaft is

---

[35] Hallis, F. (1930). *Corporate Personality – A Study in Jurisprudence*. Oxford, UK: Oxford University Press, p. 140.

[36] This term lives on in Genossenschaft (eG), a registered cooperative society under German law. A limited liability company is known as a Gesellschaft mit beschränker Haftung (GmbH) and a public corporation is an Aktiengesellschaft (AG).

rooted in rational choice.[37] Management scholars Rob Goffee and Gareth Jones based their framework for analysing corporate culture, which they analyse in terms of "sociability" and "solidarity" on the concepts of Gemeinschaft and Gesellschaft.[38]

Gierke's Genossenschaftstheorie takes as its starting point an old Germanic concept of die Gesammte Hand (the group hand).[39] The corporate organisation is the result of a creative act, giving rise to a living force. The corporate person that is created by this act is a real person (reale Gesammte Hand). The granting of legal status through incorporation is a declaratory action, not a foundational act: the role of the state is to recognise, not to create. Nor does the foundation of a corporate body necessarily involve any agreement between individuals: it is a unilateral act that has no parallel in private law. Because of this organic act of creation, it makes sense to say that the corporation has a spirit or will of its own. It is meaningful to talk about a corporation's personality. This also means that the corporation is a moral entity, a bearer of moral values, rights and obligations separate from those of its constituent members. The Genossenschaft is a real organic entity, with a separate life force, from which it derives its personality and moral stature.

Arguments between corporate realists and anti-realists continued through the early and middle parts of the twentieth century. Gierke became a major influence on other Continental European jurists, as well as English scholars F.W. Maitland (a jurist), Sir Ernest Barker (a political scientist), and Harold Laski (a sociologist, economist, and socialist political theorist), but fiction theory proved difficult to dislodge from legal thinking. In 1932, Max Radin, an American legal scholar, wrote in the Columbia Law Review of "The Endless Problem of Corporate Personality", indicating the level of on-going disagreement.[40]

[37] Weber, M. (1956 | 1978). *Economy and Society*. Los Angeles, CA: University of California Press. See especially Part 2, Chapter II "The economic relationships of organised groups". pp. 339–354.

[38] Goffee, R., & Jones, G. (1998). *The Character of the Corporation*. London: HarperCollins.

[39] In this section I am very indebted to Hallis (1930) Part III, Chapter 1, entitled "The Theory of the Reality of Corporate Persons: Otto von Gierke", p. 137–165.

[40] Radin, M. (1932). The Endless Problem of Corporate Personality. *Columbia Law Review, 32*(4), 643–667.

LAW AND ECONOMICS AFTER 1980

Between 1930 and 1980 Anglo-American legal scholars did little theorising about the nature of the corporation.[41] They may have been affected by
an influential article published by the philosopher John Dewey
(1859–1952), a pragmatist of the American school, who argued that legal
questions should be resolved by assessing the consequences of different
practical solutions (an approach to legal scholarship known as "conse-
quentialism"), not by deduction from high theory. Mainstream economists continued to show little interest in the theory of the firm. Large
managerial corporations, as described by Berle and Means in 1932, continued to dominate the business scene, meaning that legal fiction, concession, and aggregate theories of the corporation seemed increasingly
implausible. However, all this was to change after 1976 with the publication of Jensen and Meckling's article on the theory of the firm. Jensen and
Meckling's article took the economic world by storm: it is now one of the
most highly cited articles in both the management and economics canons.
Ignoring 200 years of legal scholarship, it confidently asserted: "it is
important to recognise that most organisations are simply legal fictions
which serve as a nexus of contracting relationships among individuals".[42]
By the early 1980s, in particular, because of the growth of the "law and
economics" movement, the "legal fiction "and "nexus of contracts" view
of the firm, combining fiction theory and aggregate theory, had come to
dominate legal scholarship.

The law and economics movement has its origins in the works of Ronald
Coase (especially his 1960 paper entitled "The problem of social cost")
and of Guido Calabrasi (especially his 1961 paper entitled "Some thoughts
on risk distribution and the law of torts"). It became established in the
1960s at the University of Chicago where there was both a well-regarded
law school and economics department. Although trained as an economist,
Coase became a faculty member of the University of Chicago Law School
in 1964 and remained there in various capacities for the rest of his long life
(he died in 2013 at the age of 102). He was one of the first editors of the
Journal of Law and Economics, also based in Chicago.

---

[41] Phillips, M. (1994). Reappraising the real entity theory of the corporation. *Florida State
University Law Review*, 21(4), 1061–1123.
[42] Jensen, M. & Meckling, W. (1976) p. 310.

The law and economics movement's fundamental method is to apply microeconomic techniques to analyse the law. The movement's main insight is clearly stated by Frank Easterbrook and Daniel Fischel in *The Economic Structure of Corporate Law:* "The normative thesis…is that corporate law should contain the terms people would have negotiated, were the costs of negotiating at arm's length for every contingency sufficiently low. The positive thesis is that corporate law almost always conforms to this model."[43] In other words, at least according to Easterbrook and Fischel, legal rules both should and do satisfy principles of economic efficiency. An evolutionary process of selection will generate efficient legal rules. This is particularly true in common law countries where the law is partly judge made from the legal reasoning in decided cases. Many of the movement's original ideas were summarised by Richard Posner, a federal judge as well as a senior lecturer at the University of Chicago Law School, in his magisterial work *Economic Analysis of Law*, first published in 1973.

One consequence of the rise of the law and economics movement and the eminence of its practitioners was that certain axioms of economics became embedded in interpretation of the law. This included the "legal fiction" and "nexus of contracts" conceptualisation of the corporate form, as well as the "shareholder primacy" view of the corporate objective. The sole responsibility of business, according to Milton Friedman in an article published in *The New Times Magazine* in 1970, is acting within the law and without deception and fraud, to increase profits. The corporate objective was "shareholder value maximisation".

Thus began, in or around 1980, a new phase of the corporate era, described by sociologist Gerald Davis as "The New Financial Capitalism".[44] "Financialisation", as some commentators have described it, had a number of consequences for large public corporations. It no longer became accepted wisdom that companies should spread their risk across a number of unconnected business lines: investors could make these decisions for themselves. Conglomerates ceased to be in vogue. Investment banks and private equity houses recognised that value could be realised for shareholders (and themselves) by orchestrating "bust-up" takeovers, buying

---

[43] Easterbrook, F., & Fischel, D. (1948 |1991). *The Economic Structure of Corporate Law.* Cambridge, Mass: Harvard University Press, p. 15.

[44] Davis, G. (2009). *Managed By Markets: How Finance Re-shaped America.* Oxford & New York: Oxford University Press.

public corporations at a premium to the current share price, demerging their businesses, and selling the various components at a profit. The sum of the parts was, in many cases, seen to be more valuable than the aggregate whole.

Not all commentators accepted the new conventional wisdom. Stakeholder theory questions agency's theory's central concept of shareholder primacy, arguing instead that shareholders are only one of a number of important interest groups; other stakeholders include employees, customers, suppliers, and local communities.[45] As discussed in Chap. 2, Margaret Blair (an economist) and Lynn Stout (a legal scholar) advanced the "team production theory of corporate law", perhaps the most notable example of corporate law stakeholder theory.[46] Blair and Stout propose that public corporations comprise teams of people making specific investments in the form of both financial and human capital who enter into a complex agreement to work together for mutual gain under a "mediating hierarchy". They developed their arguments in a series of articles and books.[47] Implicit throughout, and in stakeholder theories generally, is that there is more to a public corporation than a "legal fiction" and "nexus of contracts".

Arguments for and Against Corporate Realism

I turn now to the technical arguments which support corporate realism. The first argument comes from Gierke, that we cannot make sense of the world unless we conceptualise corporations as real. On this basis Max Radin, in his aforementioned article, "The Endless Problem of Corporate Personality", comes down in favour of corporate realism. In support of Gierke's argument, he quotes William of Occam's rule: "entia non sunt

---

[45] Freeman, R. (1984 | 2010). *Strategic Management: A Stakeholder Approach.* Cambridge, UK: Cambridge University Press.

[46] Blair, M., & Stout, L. (1999). A team production theory of corporate law. *Virginia Law Review, 85*(2), 247–328.

[47] Other works include: Blair, M. (1995). *Ownership and Control: Rethinking Corporate Governance for the 21st Century.* Washington, DC: The Brookings Institution; Blair, M. (1996). *Wealth Creation and Wealth Sharing.* Washington, DC: The Brookings Institution; Stout, L. (2012). *The Shareholder Value Myth: How Putting Shareholders First Harms Investors, Corporations, and the Public.* San Francisco, CA: Berrett-Koechler Publishers, Inc.

multiplicanda prater necessitatem" (do not multiply entities unnecessarily—the problem-solving principle known as "Occam's razor" or "the law of parsimony"); Radin says it is necessary to recognise real corporate entities if we are to develop a proper understanding of the world. Related to this is an argument against the combination of legal fiction and concession theory: if corporations are legal fictions, then why is the state not also regarded as a legal fiction, and if the state is also legal fiction, how can it make sense for the state (not a real person) to attribute personhood to a corporation? In the UK sovereignty is vested in Parliament. In the US sovereignty is vested for some purposes in the Federal Government and for other purposes in individual States. These are themselves "bodies corporate". The only way to avoid this argument would be if sovereignty were to be invested in an individual, for example, in a king, queen, or president. Even then, legal scholars would probably argue that for these purposes the king, queen, or president was a "corporation sole", that the office is not the same as the office-holder.[48]

The second argument is that the firm is more than the sum of its parts.[49] In an article published in 1905, Jethro Brown, an Australian legal scholar and pupil of F.W. Maitland, considered whether corporate personality is merely a useful legal device, or whether the law reflects certain real deep structures, which exist in society wholly apart from their legal recognition. In support of the latter, he writes:

> Whenever men act in common they inevitably tend to develop a spirit which is something different from themselves taken singly or in sum. No one who has had any experience as a member of a governing body, for example, can be ignorant of the fact that the decisions of such a body, even when they are unanimous, are often inexplicable if regarded from the point of view of the

---

[48] Maitland, F. W. (2003). *State, Trust and Corporation* (D. Runciman & M. Ryan Eds.). Cambridge, UK: Cambridge University Press. See Chapter 2, "The Crown as Corporation". pp. 32–51.

[49] I am heavily indebted to this and the following argument to Phillips, M. (1994). Reappraising the real entity theory of the corporation. *Florida State University Law Review*, 21(4) pp. 1061–1123. However, I depart from Phillips in a number of respects: I interpret Laski's argument as being in essence the same as Brown's, and therefore a stronger argument than Phillips gives credit for. Similarly, Machen's argument is essentially a linguistic argument, so that the simplicity of his mathematics, which Phillips rightly criticises, can be overlooked.

several characters of the individual members considered as so many units. When at a meeting of such a board a speaker begins with the statement, "I speak as a member of this board, and I say -," there will be reason to antici- pate statements or proposals which are not adequately representative of the person making them. Under the inspiration of spirit de corps, the humane will give a cruel decision, the cruel a humane.[50]

Harold Laski argued in a similar way (for an influential socialist his exam- ples, especially the last one, are rather intriguing):

> We are compelled to personalise… associations…they govern a single verb… The Bank of England is…the "little old lady of Threadneedle Street"; but no one would speak of seven distinguished merchants as a little old lady. The House of Commons is distinct from "its" members, and, no less clearly, it is not the chamber in which they meet. We talk of "its" spirit and "complexion"; a general election, so we say, changes "its" character. Eton, we know well enough, is not six hundred boys, nor collection of ancient buildings.[51]

Support for the "whole is more than the sum of the parts" argument can be found more recently in the strategic management literature. The resource-based view of the firm proposes that competitive advantage is obtained from bundles of unique resources which are available to indi- vidual firms. While these resources may be tangible (e.g., natural resources like minerals or oilfields), according to strategic management theorists they are often intangible and therefore harder to imitate. Thus, Edith Penrose postulates "bundles of human resources", Nelson and Winter "organisational routines", Jay Barney "distinctive capabilities", John Kay "organisational architecture", Prahalad and Hamel "the core competen- cies of the corporation", and Teece, Pisano & Shuen "dynamic

[50] Brown, W. J. (1905 | 2008). The personality of the corporation and the state. *Journal of Institutional Economics, 4*(2), 255–273. Originally published in *Law Quarterly Review 21*(4), 365–379.
[51] Laski, H. (1916). The personality of associations. *Harvard Law Review, 29* (4), p. 405.

capabilities".[52] All of these require organisational scale and structure that goes beyond a set of contractual relationships.[53]

Laski's argument, especially his reference to "a single verb", leads neatly to the third argument, from language. In the 1911 edition of the Harvard Law Review, Arthur Machin writes as follows: "Any group of men, at any rate any group whose membership is changing, is necessarily an entity separate and distinct from its constituent members".[54] He continues the argument in a footnote:

> This can be demonstrated mathematically. Suppose a corporation is composed of two members, a and b. Let c = the corporate entity. Now, if the corporate entity is merely the equivalent of the sum of the members, then c = a + b. Now, suppose b to assign his shares to d, then c = a + d. But this cannot be unless b is the same as d, which is absurd. Therefore, c, the corporate entity, is not equivalent to the sum of the members.[55]

[52] Penrose, E (1959 | 1995) *The Theory of the Growth of the Firm*. Oxford University Press; Nelson, R, & Winter, S. (1982) *An Evolutionary Theory of Economic Change*. Cambridge, MA. Harvard University Press; Barney, J.B. (1991). Firm resources and sustained competitive advantage. *Journal of Management*, 17 (1) pp. 99–120; Kay, J. (1993) *The Foundations of Corporate Success*. Oxford University Press, Oxford. Chapters 5–8 on distinctive capabilities pp. 63–124; Prahalad, C., & Hamel, G. (1990) The core competence of the corporation. Harvard Business Review. Vol. 68, Issue 3, pp. 79–91 Teece, David, Pisano, Gary, and Shuen, Amy. (1997) Dynamic capabilities and strategic management. *Strategic Management Journal*, 18 (7) pp. 509–533.

[53] The difference between transaction costs economics (part of the standard model) and the resource-based view can be demonstrated with some simple mathematics. Imagine two individuals, X and Y, creating value in the form of a product or service x and y, either separately (on the market) or jointly (within a single firm). The difference between value created on the market or within the firm is represented by

Market value creation: $v_m = x_1 + y_1 - t_1$

Firm value creation: $v_f = x_2 + y_2 + (x_2,y_2) - t_2$

Firms exist: (1) where $t_2 < t_1$ (Coase, Williamson etc., assuming production costs equal); and/or (2) where $[x_2 + y_2 + (x_2,y_2)] > (x_1 + y_1)$ (Penrose, Kay, Barney, Prahalad & Hamel etc., assuming transaction costs equal). The first is transaction cost economics. The second is the resource-based view of the firm.

[54] Machin, A. (1911). Corporate personality. *Harvard Law Review*, 24(4), p. 259. Machin's arguments are continued in a second article published in the following edition of Harvard Law Review, 24(5), p. 347–365.

[55] Machin 1911, p. 259, n8.

Phillips criticises this argument as both trivial and conceptually unsound mathematics, which of course prima facie it is.[56] However, if it is interpreted as an argument from language, as I believe Machin intended, then the argument becomes much stronger. Consider, for example, a scree slope on the side of a mountain comprising many thousands of stones. Rocks continuously drop from above to join the scree slope. Stones periodically fall from the bottom. Over time the composition of the scree slope changes fundamentally. Yet we still refer to it as "the scree slope". The same can be said about waterfalls and rivers—the water changes all the time.[57] Machin puts forward similar arguments for houses (which are more than their constituent bricks and mortar), schools (which are more than the pupils in attendance at a particular time; c.f., Laski's argument about Eton), and so on.

Sanford Shane, a professor of linguistics at the University of California, further develops the argument from language. He begins his paper by disagreeing with the pragmatic philosopher John Dewey, who argued that it was merely a matter of ordinary language usage to speak about institutions as if they were persons, and that this way of talking should not be taken to imply the separate existence of such institutions in the eyes of the law. "Put roughly" according to Dewey, "'person' signifies what the law makes it signify."[58] Shane argues on the contrary that ordinary language usage supports the deep structures in society, and that these are more than mere language games.

For the fourth argument, I take as a starting point a proposition which Roger Scruton attributes to Sir Ernest Barker, that: "we cannot attribute moral responsibility to a group without also attributing it to members of the group, so that the responsibility of the members exhausts the content of the corporate liability".[59] Scruton contests this, pointing out that it would only be true if the two responsibilities were the same so that attribution to the company becomes redundant. However, groups can commit offences which lie beyond the capacity of any individual. BP was held to be morally responsible by the American public for the Deepwater Horizon

---

[56] Phillips (1994) p. 1103.

[57] This is the ancient doctrine of "everything flows", made famous by Heraclitus. "No man ever steps in the same river twice" – as Plato says in the dialogue Cratylus, 401d and 402a.

[58] Dewey, J. (1926). The historic background of corporate legal personality. *Yale Law Review, 35*(6), p. 655.

[59] Scruton (1990) p. 64.

oil spill in 2010. In November 2012, BP and the US Department of Justice settled federal criminal charges, with BP pleading guilty to 11 counts of manslaughter, two misdemeanours, and a felony charge of lying to Congress. However, charges against a number of BP executives and two rig supervisors were eventually dropped. While the reputations of senior executives of BP, including the then CEO Tony Hayward, were damaged as a result of the Deepwater Horizon oil spill, it is not clear that they were held to be morally responsible for the disaster, and they were not subject to legal charges. In a way which is consistent with this, F.W. Maitland points out that if a corporation was a legal fiction it could not in any meaningful sense be said to commit torts or breaches of statutory rules, but in practice we know that a company can be held to have committed a crime of which no individual is guilty.[60]

A fifth argument is provided by Meir Dan-Cohen by way of a thought experiment which he calls: "the person-less corporation".[61] He invites us to imagine that an entrepreneur starts a manufacturing enterprise, gradually takes on staff, and eventually decides to incorporate his business as a limited company. Initially, the entrepreneur holds all the shares, but after some time he decides to raise capital by selling shares on the stock exchange, and as a result shareholdings become widely dispersed. The company continues to be highly profitable, and its managers (by this stage the founder has retired) decide to repurchase all the shares so that the company becomes an "ownerless" corporation.[62] Finally, advances in technology lead to automation of the manufacturing activity, a sharp reduction in staff levels, and computerisation of all management functions. Eventually the few remaining managers realise that there is nothing left for them to do and decide to retire, leaving a person-less, shareholder-less, corporation run by machines.

The last part of the story is of course far-fetched, as are many philosophical thought experiments, but nevertheless it is not beyond the powers of imagination. Dan-Cohen is making the point that by the end of the

---

[60] See Maitland's introduction to Gierke's *Political Theories of the Middle Age* at xxxix, cited by Hager, M. (1988). Bodies politic: The progressive history of organisational "real entity" theory. *University of Pittsburg Law Review, 50*(5), p. 589.

[61] Dan-Cohen, M. (2016). *Rights, Persons, and Organizations* (Second ed.). New Orleans: Quid Pro Books. pp. 41–43.

[62] The laws of most countries would prevent this, but the story of the person-less corporation is a thought experiment, so that some philosophical licence is permitted.

story it makes no sense to argue that the person-less corporation is a legal fiction; ontologically speaking, it appears to have a real existence, and yet it has no shareholders and employs no natural persons. Its ontological status may have changed along the way; in the early days, it might have been difficult to draw a sharp distinction between the entrepreneur and his business. However, by the end of the story it is clear that the corporation must be thought of as a real entity.

CORPORATE REALISM AND THE PUBLIC CORPORATION

I have put forward the case for corporate realism. There are arguments in favour of fiction theory that I have not covered, and some of the arguments set out above for corporate realism are less convincing than others.[63] Lawyers may dislike metaphysical arguments such as those advanced by Otto Gierke—as Harold Laski puts it: "English lawyers, it is said, have a dislike of abstractions"[64]—and the same is probably true of many economists and some management scholars. The story of the person-less corporation has also shown that the ontological status of all corporate bodies is not necessarily the same. In order to simplify the argument, I now return to this chapter's presenting issue, the ontological status of *public corporations*, and assert that the arguments in favour of regarding public corporations as real entities are overwhelming. When we think of a natural person, we think of someone (1) located in time, with a past, a future, and a memory of things gone by; (2) with an identity, such that their persona is in some way recognisable to themselves and others; (3) with a personality, which, barring calamitous accidents will remain fundamentally unaltered over the course of a lifetime; and (4) with a physical presence which is observable and recognisable, although of course it changes gradually over

---

[63] Nor are fiction theory and corporate realism necessarily the only possibilities. Jeroen Veldman and Hugh Willmott (2017) argue that binary distinctions between real entities and fictions are misconstrued, that the "single legal entity" (SLE) and corporation are different things, and that "the social ontology of the SLE and the modern corporation are radically contingent: they are inescapably contested; and their stabilisation and institutionalisation is the outcome of a political process". See Veldman, J., & Willmott, H. (2017). Social ontology and the modern corporation. *Cambridge Journal of Economics, 41*(5), 1489–1504.

[64] Laski (1916) p. 424.

time. I would contend that, of these features, (1) and (2) are equally true of public corporations, which have identities (we are able to distinguish BP from Shell, and Apple from Microsoft), corporate memories (both in the sense of what is held in corporate archives and in the minds of long-standing employees).[65] Some readers may find it difficult to think of a corporation as having (3) a personality, but will nevertheless be able to accept that a company has a corporate culture. As the famous organisational scholar Edgar Schein says: "culture is to a group what personality or character is to an individual".[66] Schein defines culture in terms of shared beliefs, values, mental models, rituals, symbols, and history, all of which have close analogues in individual personality. Like personality, organisational culture will evolve over time but will only rarely alter fundamentally. In terms of (4), it is true to say that public corporations typically have some kind of physical presence, in terms of buildings, plants, construction sites, representative people, websites, and so on, even if this is rather different from what is observable when it comes to natural persons. Ian Clarke has explored at length the relationship between place and identity in multinational corporations, focusing particularly on the case of ICI.[67] We can see the Microsoft campus in Redmond, Washington; ICI used to be identified with its Millbank Head Office in London and industrial complex on Teeside; Ford will forever be associated with the River Rouge Complex in Dearborn, Michigan.

If a further argument is needed, the French sociologist, Pierre Bourdieu's concept of *habitus* links the cognitive structures of each individual operating within an organisational field with the organisation's collective social structures. In Bourdieu's terminology, individual personality and organisational culture are *homologous*. The two concepts have shared ancestry and common evolutionary origins.[68]

---

[65] Walsh, J., & Ungson, G. (1991). Organizational memory. *Academy of Management Review,* 16(1), pp. 57–91.

[66] Schein, E. (1985 | 2004). *Organizational Culture & Leadership.* Jossey-Bass, p. 8.

[67] Clarke, I. (1985). *The Spatial Organisation of Multinational Corporations.* London & Sydney: Croom Helm.

[68] Sallaz, J., & Zavisca, J. (2007). Bourdieu in American sociology, 1980–2004. *Annual Review of Sociology, 33*(1), 21–41.

WHY IT MATTERS

Why does the ontological status of the corporation matter? Michael Phillips, a legal scholar, describes the importance of corporate real entity theory like this:

> Its acceptance would influence debates over certain legal questions and over certain issues involving the social control of corporations. In particular, if the theory is valid, the drives, interests, and obligations of the corporate group become factors that ought to be considered in some decisional contexts. More specifically, if the theory is true and if the corporate real entity assumes certain forms, corporations could have moral duties that sometimes might affect legal and policy deliberations.[69]

Jensen and Meckling, who subscribe to the fiction theory of corporations, argue on the contrary that, because the firm is not an individual (i.e., is not a natural person) it makes no sense to ask a question such as "does the firm have a social responsibility?" The question is meaningless, they say, because firms are not "persons with motivations and intentions".[70] The primary objective of this chapter is to rebut the agency theorists' assertion that public corporations are merely legal fictions. This is not really the place to examine further whether corporations have moral responsibilities that are not decomposable into the moral obligations of individual members, directors, and executives. However, I happen to believe that public corporations do have moral and social obligations as well as legal responsibilities. In the same way that corporations have identities, personalities (organisational cultures), and physical presence, so they also have unique internal decision-making structures and processes which, I would argue, satisfy the condition set by some moral philosophers that acts must be *intentional* if they are to be judged as morally good or bad. We attribute values, goals, strategies, and other apparently personal attributes to public corporations—why not also intentionality and conscience?[71] Kenneth Goodpaster, a Harvard Business School business ethics teacher, summarised the position like this: "the fact that corporations are much more like per-

---

[69] Phillips, M. (1994). Reappraising the real entity theory of the corporation. *Florida State University Law Review, 21*(4), 1061–1123, p. 1074.

[70] Jensen & Meckling (1976) p. 311.

[71] Goodpaster, K., & Matthews, J. (1982). Can a corporation have a conscience? *Harvard Business Review*, 60 (1), pp. 132–141.

sons than [machines or animals] is therefore significant... this fact is doubt-less rooted not only in the intelligibility of attributing *intentions, decisions and actions* to organisational bureaucracies, but in the intelligibility of attributing *rationality* (or lack of it) to those intentions, decisions and actions".[72]

There is a further reason why the ontological status of corporations matters, which is more pertinent to the subject matter of this book. Supporters of the standard model of the firm would argue as follows: (1) if corporations do not exist (in the sense that we are supposed to look through them to the underlying contracts); and (2) if stockholders own corporations; then (3) the standard principal-agent model makes sense— shareholders (principals) contract with executives (agents) to manage the corporation on their behalf. I have argued, however, that: (4) public corporations are real entities, and (5) shareholders own shares, not companies, so neither of the two premises (1) and (2) above is correct. This has serious consequences for agency theory.

The alternative model of agency that I am advancing goes like this: (1) senior executives contract with corporations, which have legal, economic, and ethical capacity; (2) in the case, particularly, of directors, including executive directors, these contracts come with fiduciary responsibilities— corporate managers have trustee-like duties and obligations, requiring them to serve in the best interests of the company and of those who, continuing the parallels with trust law, are the companies' beneficiaries, including shareholders, employees, customers, and the public generally. In 1929, Owen D. Young (1874–1962) an American industrialist, lawyer, and diplomat, who was president and chairman of General Electric between 1922 and 1939, described his responsibilities thus:

> It makes a great difference in my attitude toward my job as an executive officer of the General Electric Company whether I am a trustee of the insti-

---

[72] Goodpaster, K. (1983) The concept of corporate responsibility. *Journal of Business Ethics*, 2 (1), p. 15. There are of course philosophers who argue that corporate moral agency is a fallacy, notably Velasquez (1985). There is in fact an extensive literature on the subject, originating with French (1979). Literature reviews supporting opposing positions are provided by Moore (1999) and Rönnegard (2015). Winkler (2018) explains some of the undesirable moral consequences of corporate realism in his description of how corporations in America became "persons" entitled to constitutional rights which they used to counteract the activities of regulators.

tution or an attorney[73] for the investor. If I am a trustee, who are the beneficiaries of the trust? To whom do I owe my obligations? ...

Now, I conceive my trust first to be to see to it that the capital which is put into my concern is safe, honestly and wisely used, and paid a fair rate of return. Otherwise we cannot get capital. The worker will have no tools. Second, that the people who put their labour and lives into this concern get fair wages, continuity of employment, and a recognition of their right to their jobs where they have educated themselves to highly skilled and specialized work. Third, that the customers get a product which is as represented and that the price is such as is consistent with the obligations to the people who put their capital and labour in. Last, that the public has a concern functioning in the public interest and performing its duties as a great and good citizen should.

I think what is right in business is influenced very largely by the growing sense of trusteeship which I have described. One no longer feels the obligation to take from labour for the benefit of capital, not to take from the public for the benefit of both, but rather to administer wisely and fairly in the interest of all.[74]

The alternative theory of agency is beginning to take shape. Corporate managers,[75] as agents of the company, with trustee-like responsibilities, are appointed to look after the interests of the corporation and its various classes of beneficiaries—shareholders, employees, and so on. They have legal and ethical obligations to the aforementioned, and to some extent society as a whole. Their primary objective is to maximise the total market value of the firm (the "total firm value maximisation principle" described in Chap. 2). They are responsible for ensuring that value is shared in an

---

[73] It is perhaps natural that Young, who first trained and practised as a lawyer, frames the antithesis in terms of the relationship between attorney and client, which is, of course, another paradigmatic example of a principal-agent relationship.

[74] Quoted in E. Merrick Dodd, Jr. (1932). For whom are corporate managers trustees? *Harvard Law Review*, 45 (7), pp. 1154–1155.

[75] By talking about "corporate managers" I skate over here the important legal distinction between directors (executive and non-executive), on the one hand, and senior corporate executives (e.g., those who sit on the management committee, executive committee, or similar) on the other hand. While the former, in English and American state law, certainly have fiduciary duties, the position of the latter is less clear. This becomes a normative issue, for in the alternative theory of agency it is clearly important that everyone who is a "corporate manager" should bear duties and obligations over and above those that are typically represented in a standard principal-agent relationship.

appropriate way, between shareholders, who receive value in the form of dividends and capital gains, employees, through wages and incentives, and so on.

One of the benefits of the standard model is that, on the face of it, it solves the problem of allocating value: employees receive wages, debt providers receive interest, customers acquire products at prices that are at or below the maximum amount they would be willing to pay,[76] and so on; shareholders, as residuary beneficiaries, get whatever is left, as dividends and capital gains. One problem with the alternative model sketched out above is that allocation of value is not so straightforward. How to allocate economic benefits in a public corporation, with large numbers of potential beneficiaries participating in a common pool of value, becomes a kind of collective action problem, to which I turn in the next chapter.

Further Reading
The key readings on critical realism can be found in: Archer, M., Bhaskar, R., Collier, A., Lawson, T., & Norrie, A. (1998) *Critical Realism – Essential Readings*. Routledge. An old, but still quite readable work on the nineteenth-century German jurists is: Hallis, F. (1930). *Corporate Personality – A Study in Jurisprudence*. Oxford University Press. A short book on the history of the corporation is: Micklethwait, J., & Wooldridge, A. (2003) *The Company. A Short History of a Revolutionary Idea*. London: Weidenfeld & Nicolson. Baars, G., & Spicer, A. (2017). *The Corporation: A Critical, Multidisciplinary Handbook*. Cambridge University Press., is another valuable resource.

REFERENCES

Baars, G., & Spicer, A. (2017). *The Corporation: A Critical, Multidisciplinary Handbook*. Cambridge: Cambridge University Press.
Barney, J. (1991). Firm Resources and Sustained Competitive Advantage. *Journal of Management*, *17*(1), 99–120.
Bhaskar, R. (1975). *A Realist Theory of Science*. Leeds: Leeds Books Ltd.

---

[76] This is the way that surplus is thought about in the strategy literature. The difference between a consumer's willingness to pay (WTP) and the producing firm's input costs represents a surplus, which is divided between the producing firm (profit) and the customer (consumer surplus). Producers set prices below a consumer's in order to incentivise them to buy.

74   A. PEPPER

Bhaskar, R. (1979). *The Possibility of Naturalism – A Philosophical Critique of the Contemporary Human Sciences*. Brighton: Harvester Press.

Blair, M., & Stout, L. (1999). A Team Production Theory of Corporate Law. *Virginia Law Review, 85*(2), 247–328.

Bratton, W. (1989). The New Economic Theory of the Firm: Critical Perspectives From History. *Stanford Law Review, 41*(6), 1471–1527.

Brown, W. J. (1905| 2008). The Personality of the Corporation and the State. Journal of Institutional Economics, 4(2), 255–273.

Calabresi, G. (1961). Some Thoughts on Risk Distribution and the Law of Torts. *The Yale Law Journal, 70*(4), 499–553.

Clarke, I. (1985). *The Spatial Organisation of Multinational Corporations*. London/Sydney: Croom Helm.

Coase, R. (1960). The Problem of Social Cost. *Journal of Law and Economics, 3*(October), 1–44.

Cottrell, P. (1979). *Industrial Finance 1830–1914. The Finance and Organization of English Manufacturing Industry*. London/New York: Methuen.

Dale, R. (2004). *The First Crash – Lessons from the South Sea Bubble*. Princeton: Princeton University Press.

Dan-Cohen, M. (2016). *Rights, Persons, and Organizations* (2nd ed.). New Orleans: Quid Pro Books.

Davis, G. (2009). *Managed By Markets: How Finance Re-shaped America*. Oxford/New York: Oxford University Press.

Dewey, J. (1926). The Historic Background of Corporate Legal Personality. *Yale Law Review, 35*(6), 655–673.

Dodd, E. (1932). For Whom Are Corporate Managers Trustees? *Harvard Law Review, 45*(7), 1145–1163.

Donaldson, L. (1995). *American Anti-management Theories of Organization*. Cambridge: Cambridge University Press.

Easterbrook, F., & Fischel, D. (1948 |1991). *The Economic Structure of Corporate Law*. Cambridge, MA: Harvard University Press.

French, P. (1979). The Corporation as a Moral Person. *American Philosophical Quarterly, 16*(3), 207–215.

Ghoshal, S. (2005). Bad Management Theories Are Destroying Good Management Practices. *Academy of Management – Learning & Education, 4*(1), 75–91.

Goffee, R., & Jones, G. (1998). *The Character of the Corporation*. London: HarperCollins.

Goodpaster, K. (1983). The Concept of Corporate Responsibility. *Journal of Business Ethics, 2*(1), 1–22.

Goodpaster, K., & Matthews, J. (1982). Can a Corporation Have a Conscience? *Harvard Business Review, 60*(1), 132–141.

Hager, M. (1988). Bodies Politic: The Progressive History of Organisational "Real Entity" Theory. *University of Pittsburgh Law Review, 50*(5), 575–654.

Hallis, F. (1930). *Corporate Personality – A Study in Jurisprudence.* Oxford: Oxford University Press.

Harris, R. (2000). *Industrializing English Law. Entrepreneurship and Business Organization, 1720–1844.* Cambridge: Cambridge University Press.

Jones, G. (2002). Unilever in the United States, 1945–1980. *Business History Review, 76*(3), 435–478.

Kay, J. (1993). *The Foundations of Corporate Success.* Oxford: Oxford University Press.

Laski, H. (1916). The Personality of Associations. *Harvard Law Review, 29*(4), 404–426.

Lawson, T. (2015a). A Conception of Social Ontology. In S. Pratten (Ed.), *Social Ontology and Modern Economics* (pp. 19–52). London/New York: Routledge.

Lawson, T. (2015b). The Nature of the Firm and Peculiarities of the Corporation. *Cambridge Journal of Economics, 39*(1), 1–32.

Locke, J. (1690 | 1960). *An Essay Concerning Human Understanding.* Glasgow: Collins.

Machin, A. (1911). Corporate Personality. *Harvard Law Review, 24*(4), 253–267.

Maitland, F. W. (1911 | 2003). *State, Trust and Corporation.* Ed. D. Runciman & M. Ryan. Cambridge: Cambridge University Press.

Micklethwait, J., & Wooldridge, A. (2003). *The Company. A Short History of a Revolutionary Idea.* London: Weidenfeld & Nicolson.

Moore, G. (1999). Corporate Moral Agency: Review and Implications. *Journal of Business Ethics, 21*(1), 329–343.

Nelson, R., & Winter, S. (1982). *An Evolutionary Theory of Economic Change.* Cambridge, MA: Harvard University Press.

Penrose, E. (1957 | 2009). *The Theory of the Growth of the Firm.* Oxford: Oxford University Press.

Perrow, C. (1986). *Complex Organizations.* New York: Random House.

Phillips, M. (1994). Reappraising the Real Entity Theory of the Corporation. *Florida State University Law Review, 21*(4), 1061–1123.

Posner, R. (1973 | 2014). *Economic Analysis of Law.* New York: Wolters Kluwer.

Prahalad, C., & Hamel, G. (1990). The Core Competence of the Corporation. *Harvard Business Review, 68*(3), 79–91.

Radin, M. (1932). The Endless Problem of Corporate Personality. *Columbia Law Review, 32*(4), 643–667.

Rönnegard, D. (2015). *The Fallacy of Corporate Moral Agency.* Heidelberg: Springer.

Sallaz, J., & Zavisca, J. (2007). Bourdieu in American Sociology, 1980–2004. *Annual Review of Sociology, 33*(1), 21–41.

Schein, E. (1985 | 2004). *Organizational Culture & Leadership.* San Francisco: Jossey-Bass.

Scruton, R. (1990). Gierke and the Corporate Person. In *The Philosopher on Dover Beach.* New York: St Martin's Press.

Sloan, A. (1964). *My Years with General Motors*. Garden City: Doubleday & Company, Inc.

Teece, D., Pisano, G., & Shuen, A. (1997). Dynamic Capabilities and Strategic Management. *Strategic Management Journal, 18*(7), 509.

Velasquez, M. (1985). Why Corporations Are Not Morally Responsible for Anything They Do. In J. Desjardins & J. McCall (Eds.), *Contemporary Issues in Business Ethics*. California: Wadsworth.

Veldman, J., & Willmott, H. (2017). Social Ontology and the Modern Corporation. *Cambridge Journal of Economics, 41*(5), 1489–1504.

Walsh, J., & Ungson, G. (1991). Organizational Memory. *Academy of Management Review, 16*(1), 57–91.

Weber, M. (1956 | 1978). *Economy and Society*. Los Angeles: University of California Press.

Willman, P. (2014). *Understanding Management – the Social Science Foundations*. Oxford: Oxford University Press.

Wilson, C. (1954). *The History of Unilever*. London: Cassell.

Winkler, A. (2018). *We the Corporations: How Amercian Business Won Their Civil Rights*. New York: Liveright Publishing Corporation.

CHAPTER 4

Executive Pay as a Collective Action Problem

Abstract This chapter takes as its starting point Mancur Olson's assertion in *The Logic of Collective Action* that his theory of group size and group behaviour has implications for the governance of companies. It explains why shareholders of public corporations are unlikely to solve executive pay problems because of a collective action problem, and how ideas about the governance of common pool resources have implications for the design of corporate governance mechanisms. A study of the FTSE 100 is used to illustrate the points raised.

Keywords Collective action • Corporate governance • Common pool resources

INTRODUCTION

The UK is widely regarded as having one of the most robust company law and corporate governance regimes in the world,[1] and it was one of the first countries to introduce "say on pay" provisions for shareholders. Yet investors are often reluctant to vote against executive pay proposals—collective action problems mean it is difficult to bring together a shareholder alliance

---

[1] Charkham, J. (1995). *Keeping Good Company: A Study of Corporate Governance in Five Countries.* Oxford, UK: Oxford University Press.

© The Author(s) 2019   77
A. Pepper, *Agency Theory and Executive Pay,*
https://doi.org/10.1007/978-3-319-99969-2_4

sufficient to reject a directors' remuneration report. Government and the press urge boards to take more responsibility for moderating pay claims, yet non-executive directors may believe that the costs of challenging claims outweigh the benefits; they may wish to avoid conflict over the pay of executives who are fellow directors on unitary boards, and may feel compromised because they are or were themselves at one time executives of other large companies.

In his book, *The Logic of Collective Action*, Mancur Olson argues that his theory of group size and group behaviour has implications for the governance of companies. He writes,

> The autonomy of management in the large modern corporation, with thousands of stockholders, and the subordination of management in the corporation owned by a small number of stockholders, may also illustrate the special difficulties of the large group. The fact that management tends to control the large corporation and is able, on occasion, to further its own interest at the expense of the stockholders, is surprising, since the common stockholders have the legal power to discharge the management at their pleasure, and since they have, as a group, also an incentive to do so, if the management is running the corporation partly or wholly in the interest of managers. Why, then, do not the stockholders exercise their power? They do not because, in a large corporation, with thousands of stockholders, any effort the typical stockholder makes to oust the management will probably be unsuccessful; and even if the stockholder should be successful, most of the returns in the form of higher dividends and stock prices will go to the rest of the stockholders, since the typical stockholder owns only a trifling percentage of outstanding stock. The income of the corporation is a collective good to the stockholders, and the stockholder who holds only a minute percentage of the total stock, like any member of a latent group, has no incentives to work in the group interest. Specifically, he has no incentive to challenge the management of the company.[2]

In this chapter I shall describe how Olson's theory of groups and organisations provides insights that can be used to enhance the standard model and explains how excessive executive pay can be modelled as a collective action problem. In doing so I retain, for the time being, the standard economic assumptions of profit-seeking corporations, rational, rent-seeking

[2] Olson, M. (1965|1971). *The Logic of Collective Action – Public Goods and the Theory of Groups*. Cambridge, Mass: Harvard University Press, p. 55.

principals and agents, and no non-pecuniary agent motivation. A study of public companies that make up the FTSE 100 index in the UK is used to help build the theory. For the time being, I also work within the parameters of the standard model, which assumes that shareholders have the rights to all residual profits, that is, those calculated after deducting all factor inputs. The chapter concludes by proposing various ways in which the standard model can be improved upon, including comments on how stakeholders other than shareholders might share in the common pool, building on the ideas put forward at the end of the previous chapter.

MODELLING AGENCY COSTS AS A COLLECTIVE ACTION PROBLEM

Olson argues that the autonomy of managers in large modern corporations is a specific example of his general theory of groups and organisations. The fact that executives exercise management and control over large corporations and are able, on occasions, to further their own interests at the expense of shareholders might be recognised as a collective action problem.[3] There is a sense in which the earnings of a corporation are a collective good to stockholders, so that a shareholder owning a small percentage of total stock is like any member of what Olson calls a "latent group",[4] with no incentive to challenge the management of the company as the costs in doing so are likely to outweigh the potential benefits. Russell Hardin points out that collective action problems often relate to the elimination of a cost, which constitutes a good to those who would otherwise bear that cost.[5] Joseph Heath describes a large corporation as a "quasi-public good" to its members. They all derive benefits from the corporation, but individual self-interested action will not secure those benefits. In order to produce and sustain these quasi-public goods, as Heath says: "it is necessary to overcome a complex set of collective action problems."[6]

Olson provides a typology of groups, which he describes as being "privileged", "intermediate", or "latent". In a privileged group the benefits of

[3] Heath, J. (2014). *Morality, Competition, and the Firm*. NY, USA: Oxford University Press.

[4] Olson (1965|1971) p. 50.

[5] Hardin (1982|2013). *Collective Action*. London and New York: Routledge.

[6] Heath (2014) p. 51.

action are likely to exceed the cost for at least some of members of the group so that, other things being equal, collective action is likely to succeed. In a latent group the cost of action is likely to exceed the benefits for all group members, so that, other things being equal, the action is likely to fail. Small groups are typically privileged; large groups are typically latent; intermediate groups may behave like privileged or latent groups depending on whether coordination, benefits-sharing, and cost-sharing are or are not possible in practice. Olson offers no numerical guidance as to what constitutes small, intermediate, or large group sizes. His main conclusion, that latent groups will fail, is modified in certain circumstances. According to his "by-product" theory, groups may selectively offer private goods on favourable terms to members who agree to combine in collective action. Olson postulates that this may explain why labour unions offer healthcare, insurance, and other financial services on exclusive terms to members. According to Olson's "special interests" theory, large groups will sometimes form themselves into smaller special interest groups that are small enough to negotiate collective action arrangements among themselves. For example, business communities are typically divided into a series of industries, often containing only a small number of separate firms, in markets that tend towards oligopolistic competition. Trade associations are frequently established to represent collective industry interests in such a way that anti-trust considerations are not breached.[7]

The concepts of privileged, intermediate, and latent groups, and the by-product and special interest theories, can be used to model agency costs arising in a public corporation as a collective action problem. Formally, let E be the set of members of the executive committee of firm F, comprising n individuals indexed by $e \in [1,\dots,n]$; let D be the set of members of the board of directors of F, comprising n individuals indexed by $d \in [1,\dots,n]$; let S be the set of shareholders of F, comprising n individuals, funds, and companies, indexed by $s \in [1,\dots,n]$; and let $x = x_1,\dots,x_n$ denote the vector of financial and non-financial benefits receivable by e, d, and s in respect of their involvement with F. The utility function of executive $e \in [1,\dots,n]$ is given by

$$U_e(x) = w_e - \varepsilon_e \qquad (4.1)$$

---

[7] Levenstein, M., & Suslow, V. (2006). What determines cartel success? *Journal of Economic Literature, 46*(1), pp. 43–95.

where w_e is the executive's financial and non-financial reward received for working for F and ε_e is the executive's cost in terms of effort. The utility function of shareholder $s \in [1,...,n]$ is given by

$$U_s(x) = y_s - g_s \qquad (4.2)$$

where y_s is the financial return which s receives from F and g_s is the governance cost which s incurs in respect of the investment in F. The return y_s represents an individual stockholder's proportionate share of the total profits Y of F. In the case of minority shareholders, y_s is delivered in the form of dividends and capital gains. Y is calculated after deducting governance costs borne by F as well any rents (i.e., excessive executive compensation) paid to executives. Executive rents are given by

$$R = \sum_{e=1}^{n} \left(w_e - w_e^*\right) \qquad (4.3)$$

where w_e^* represents the financial and non-financial rewards which would be payable to its senior executives by F in the absence of any agency problems.

Olson says that the income of a corporation is a collective good for stockholders. A corollary of this is that agency costs relating to F represent a potential asset to S to which, in Olson's terminology, the logic of collective action applies. Executive rents represent a collective good to shareholders because a reduction in R would result in an increase in Y. To avoid confusion over signs I define this as $|R|$, representing executive rents embedded in F which are potentially recoverable. The proportionate share of $|R|$ due to s is represented by p_s. Eq. (4.2) can therefore be rewritten:

$$U_s(x) = y_s - g_s + p_s|R| \qquad (4.4)$$

The proportion p_s is calculated by dividing the number of shares held by s by the total number of equivalent shares issued by

F, so that $p_s = {}^{S_n}\!\big/\!{}_{S_N}$

According to Olson's logic, if S_N is large, then it is less likely that $|R|$ will be recoverable, because it is more likely that governance costs will exceed the proportion of agency costs recoverable by any one shareholder, that is, formally, $g_s > p_s|R|$. S will be, in Olson's terms, a "latent group". A latent group is distinguished by the fact that, unless one member takes the lead in providing the collective good, no other member will be significantly affected and therefore have any reason to act. Nevertheless, as Olson and others have pointed out, collective action can still occur in latent groups. Three of the possible reasons for this are relevant here. First, if for any shareholder s, s_n increases at a faster rate than any increase in S_N, then for that shareholder it is possible that at some point $p_s|R| > g_s$, making it worthwhile for s to try to solve the collective action problem and to reduce excessive executive compensation. Other members of S would benefit as free-riders from the actions of this leading shareholder without directly incurring any further governance costs themselves. Secondly, an individual shareholder might obtain ancillary benefits from leading a collective action to recover executive rents, for example, by enhancing its reputation as an active investor, thus making it worthwhile for s to pursue an action on behalf of all members of S, despite that fact that on an individual basis $g_s > p_s|R|$. Thirdly, a group of shareholders might agree to work together so that on a collective action basis the potential reduction in their share of agency costs might exceed their direct governance costs. I describe the combination of these factors as the "principal force"[8] or α. Therefore, Eq. (4.2) can be revised as follows:

$$U_s(x) = y_s - g_s + \alpha\, p_s|R| \qquad\qquad (4.5)$$

where $\alpha > 1$ if any combination of the three factors mentioned above applies; otherwise $0 \leq \alpha \leq 1$.

The last part of the model considers the position of the non-executive directors of F. The utility function of director $d \in [1,\ldots,n]$ is conventionally given by:

$$U_d(x) = w_d - \varepsilon_d \qquad\qquad (4.6)$$

where w_d is the non-executive director's financial compensation for serving F and ε_d is the director's effort cost. Some directors have a powerful

---

[8] The force which principals apply upon agents.

sense of their fiduciary responsibilities and professional ethics that defy conventional economic analysis, but can nevertheless be modelled by incorporating an additional factor into their utility functions. Directors are also subject to reputational effects. A director of good reputation can expect to obtain a portfolio of other high-status non-executive director-ships.[9] However, they are also subject to a set of onerous legal and regulatory obligations. In the model these moral, reputational, and legal effects are together combined in a factor that I call β, or the "fiduciary force". This is a coefficient that is applied to the first part of the director's utility function. Thus Eq. (4.6) is rewritten as follows:

$$U_d(x) = \beta w_d - \varepsilon_d \qquad (4.7)$$

where β can take various values. If $\beta > 1$, then the fiduciary force enhances the director's utility function. If $\beta < 1$, then the director's reputation is undermined and their utility function is correspondingly diminished.[10]

Study of UK FTSE 100

The Financial Times Stock Exchange 100 index is a share index of the 100 largest companies by market capitalisation listed on the London Stock Exchange. The index is maintained by the FTSE Group, a wholly owned subsidiary of the London Stock Exchange. Constituent companies must have a full listing on the London Stock Exchange, with sterling- or euro-denominated prices on the Stock Exchange's electronic trading service. They must also meet certain requirements regarding a free float and the liquidity of their shares. On the date the case study was prepared (December

---

[9] Negative reputational effects are also possible. In the spring of 2016, Alison Carnwath of Barclays, Dame Ann Dowling of BP, Judy Sprieser of Reckitt Benckiser, Sir John Hood of WPP, and Melanie Gee of Weir Group were all cited as remuneration committee chairs whose reputations have been damaged as a result of shareholder opposition to executive pay awards. Sources: Financial Times. Executive pay committee chiefs in the hot seat (May 3, 2016). The Times. Boardroom pay is off the scale and shareholder revolts will not reel it back (May 4, 2016).

[10] These equations are repeated at the funds level, creating a further set of agency and fiduciary relationships, where retail investors are principals and individual investment managers are agents. Depending on the structure of the relevant funds, these relationships may be mediated by non-executive fund directors.

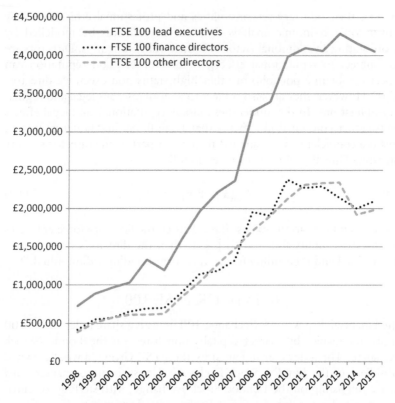

Fig. 4.1 Executive pay in the FTSE 100 in the period 2000–2015. (Based on data published by Income Data Services)

31, 2015) the largest company in the index was Royal Dutch Shell, with a market capitalisation of £160.1 billion. The total market capitalisation of companies in the index was £1.8 trillion, 73.2% of this value being represented by the top 35 companies and 85.1% by the top 50.

UK executive remuneration has escalated in the past two decades, with average chief executive pay in FTSE 100 companies reaching £4,284,000 in 2015, 171 times the average wage of employees (source: Income Data Services and Office for National Statistics)—see Fig. 4.1.

Section 439 of the UK Companies Act 2006 mandates an annual vote on directors' pay, although these "say on pay provisions" (broadly correspond to equivalent provisions introduced in the US in 2010 by Section

951 of the Dodd-Frank Reform Act) are not binding on company.[11] Shareholders voted against FTSE 100 companies' pay proposals on five occasions between 2009 and 2013, one of the largest revolts being in 2012 when nearly 60% rejected the £6.8m annual pay package of WPP chief executive Sir Martin Sorrell at the advertising agency's annual general meeting. In April 2016 investors voted against BP's remuneration report, with 59% of proxy votes cast going against the company's decision to pay its CEO Bob Dudley nearly US$20m for 2015, a year in which the company ran up a US$5.2 billion loss. Andrew Tyrie, who was then chairman of the UK parliament's influential treasury committee, urged investors to maintain their stand against excessive pay for corporate bosses following the BP vote. *The Financial Times* reported that BP's board was facing pressure from large institutional shareholders to remove Dame Ann Dowling, who chaired its remuneration committee. The leader column in the FT urged company boards to take responsibility for limiting the quantum of executive pay.[12] Yet the typical structure of shareholdings in UK public companies makes this difficult to do for the reasons explained earlier in this chapter.

Under UK company law and stock exchange rules, any investor with a direct or indirect shareholding commanding 3% or more of the voting rights in a UK public listed company is required to disclose this to the company concerned. A company is required to identify in its annual report and accounts all investors owning 3% or more of its shares at the balance sheet date. Members with shareholdings representing at least 5% of a public listed company's total voting rights can require the directors to call a general meeting and requisition the circulation of a statement regarding a proposed resolution. More than 50% of shareholders voting can pass an ordinary resolution at a general meeting, such resolutions being required, for example, to approve a related party or large transaction. Approval by 75% of shareholders voting is necessary for a special resolution, required, for example, by a public listed company to agree to a major transaction.

---

[11] The Enterprise Regulatory Reform Act 2013 introduced forward-looking provisions requiring a company to obtain shareholder approval every three years for its directors' remuneration policy. This is a binding vote, but it places a lesser obligation on the board than having to obtain approval for actual amounts paid.

[12] Financial Times: BP investors revolt over chief Bob Dudley's 20% pay rise (April 14, 2016); Boards are responsible for limiting pay excess (April 18, 2016); Tyrie adds support to revolt on excessive pay (April 18, 2016).

More than 25% of shareholders voting can block a special resolution. Investors can appoint proxies to vote on their behalf. Proxies can often play an important part in corporate governance and shareholder activism by collecting mandates and voting en bloc. A number of proxy advisory groups, including Institutional Shareholder Services (ISS), Glass, Lewis & Co., and Pensions Investment Research Consultants, issue general guidance on corporate governance and executive compensation; in some cases, they will do this by making specific voting recommendations.

A breakdown of holdings at various significant levels for each of the top 35, 50, and 100 companies in the FTSE 100 index on December 31, 2015 is set out in Table 4.1.[13] Six companies have been excluded because they had a single dominant investor.[14] On average 67.2% of shares held by the 100 largest shareholders in each company are held on behalf of retail investors by institutional shareholders (banks, insurance companies, mutual funds, pension funds, private equity firms, and other financial investors), 8.4% by trade investors, 9.0% by government (e.g., sovereign wealth funds), and only 6.1% directly by individual investors. The top 100 shareholders, on average, hold in aggregate 68.8% of the total share capital. In a typical company a small number of institutional shareholders (median = 5) have holdings of between 3% and 6% of company's share capital, five investors might control more than 25% of the voting rights, 19 might control 50% of the votes, and 82 might control 75% or more.

What is apparent from this analysis is that, except in the small number of cases that have been identified, there is no single dominant shareholder. Even though, on average, five institutional investors control 26.5% of a company's shares, sufficient to block a special resolution, this does not constitute a significant level of control, even if the five could be persuaded to act in concert. It takes on average 19 institutional investors to control 50% of the votes and 82 to control 75%. This is too large a number to make it likely that coalitions to vote down executive pay proposals will

---

[13] The data on which the analysis is based was obtained from Orbis http://www.bvdinfo.com/en-us/our-products/company-information/international-products/orbis

[14] The companies excluded from the analysis were TUI Group, Fresnillo, Schroders, Hargreaves Lansdown, Merlin Entertainments, and Sports Direct. The UK Listing Rules require a shareholder or shareholder group who could exercise 30% or more of the voting rights of a company to enter into a Relationship Agreement with the company which guarantees certain independence provisions designed to protect the rights of other shareholders.

Table 4.1 Analysis of shareholdings in FTSE 100 companies on December 31, 2015

	3% shareholders		Number of shareholders representing ≥10% of ordinary shares	Number of shareholders representing >25% of ordinary shares	Number of shareholders representing ≥50% of ordinary shares	Number of shareholders representing ≥75% of ordinary shares	Top 100 shareholders				
	Number	% of total share capital					Institutional investors	Trade investors	Investment by the State	Individual investors	% of total share capital
FTSE 35											
Mean	3.7	22.7	2.6	9.4	37.0	81.4	70.5	7.7	9.2	5.9	63.5
Median	3.0	19.7	2.0	7.0	32.0	97.0	73.0	8.0	11.0	4.0	63.8
Std. Deviation	2.3	15.2	2.5	8.2	24.7	26.1	11.5	3.1	3.7	5.8	17.8
FTSE 50											
Mean	4.2	24.3	2.4	8.3	34.6	78.7	69.2	8.3	9.1	6.4	65.4
Median	4.0	21.4	2.0	7.0	32.0	97.0	71.0	9.0	11.0	6.0	67.7
Std. Deviation	2.3	14.1	2.2	7.2	24.4	26.7	10.7	3.6	3.7	5.4	15.5
FTSE 100											
Mean	5.1	26.9	2.1	7.4	28.2	72.9	67.2	8.4	9.0	6.1	68.8
Median	5.0	26.5	2.0	5.0	19.0	82.0	70.5	9.0	10.0	6.0	72.2
Std. Deviation	3.6	13.9	1.7	7.0	21.8	26.7	12.0	7.7	9.2	5.9	17.7

Table 4.2   Investors with 3% holdings in FTSE 100 companies on December 31, 2015

Investor	Number of 3% holdings
Blackrock[a]	106
Capital Group[b]	32
Legal & General[a]	32
Fidelity[b]	20
Government of Norway[c]	20
Invesco[a]	20
Standard Life[a]	19
Aberdeen Asset Management[a]	17
Franklin Resources[a]	14
AXA[a]	12
Ameriprise Financial[a]	11
Sun Life Financial[a]	11
Others (85 investors)	206
Total	520

[a]Listed company

[b]Private company

[c]Other

naturally emerge. Furthermore, many of these institutional investors are themselves listed companies—Table 4.2 provides an analysis. This means that there is an additional disincentive for those companies to vote against executive pay proposals because of the risk of reciprocal action.

The most commonly represented institutional shareholders were Blackrock (a US listed company), Legal and General (a UK listed company), and Capital Group (a US private company).[15] Blackrock has itself been criticised for overpaying its CEO and for being "too soft" on "excessive executive remuneration" at companies in which it invests. In May 2016 75,000 people signed an online petition urging Blackrock to overhaul its approach towards executive pay at the companies in which it invests.[16]

Commentators have identified a number of possible solutions to the collective action problem among investors. An article published in *The*

---

[15] Another major investor is the Norwegian sovereign wealth fund – see Chap. 5, n17.

[16] Financial Times. BlackRock slammed over too many votes for high pay (May 22, 2016). The chief executive of BlackRock is overpaid (June 5, 2016).

Review of Financial Studies in 2008 described in detail the activities of the Hermes UK Focus Fund (HUKFF) based on private information made available by Hermes, the fund manager owned by the British Telecom Pension Scheme.[17] HUKFF generated above-average returns by engaging in private interventions with companies on various matters of performance and corporate governance, including executive remuneration. HUKFF was set up by Alastair Ross-Goobey, a well-known figure in the investment industry and a notable advocate of the need for shareholders to engage with boards and push for corporate governance reforms. The fund was created "as a response to the problem of free riding in institutional activism as perceived by the BT pension fund trustees. The trustees felt that the cost of higher intensity activism could not be sufficiently internalised through the core engagement, and it was therefore necessary to over-weight the fund's position in underperforming stocks that were to be engaged more intensively".[18] However, the case of HUKFF is an example of an exception which is the rule—HUKFF was unique among institutional investors when it was founded in 1998. Its significance declined after 2009 and it was eventually sold to RWC Partners, a London-based hedge fund, in 2012.[19]

Some hedge funds also engage in shareholder activism because of concerns about corporate governance. For example, Elliott Associates successfully fought a prolonged battle with the board of Alliance Trust, a FTSE350 investment company, over financial performance and corporate governance issues, including the pay of its chief executive Katherine Garrett-Cox. They eventually secured organisational changes, including the ousting of Garrett-Cox, before selling their stake in 2017. However, activist hedge funds typically have bigger fish to fry than executive remuneration. Interventions tend to focus on changing business strategy, especially where divestment or demerger has the possibility of realising substantial short-term capital gains.

A third possibility is that proxy advisory firms such as ISS help to coordinate the actions of disparate shareholders. PwC has examined the

[17] Becht, M., Franks, J., Mayer, C., & Rossi, S. (2008). Returns to shareholder activism: evidence from a clinical study of the Hermes UK Focus Fund. *The Review of Financial Studies, 22*(8), 3093–3129.

[18] Becht et al. 2008: p. 3102.

[19] Financial Times, September 18, 2012.

outcomes of advisory votes on remuneration reports for FTSE 100 companies for the period 2015–2017. They found evidence that ISS voting recommendations do have an impact on voting outcomes, increasing a negative vote by 10–15 percentage points when advising against a resolution.[20] However, this is a relatively marginal effect and will only occasionally cause remuneration reports to be voted down. PwC also points out companies often complain that ISS follows a mechanistic approach to voting recommendations, advising against atypical remuneration plans which depart from established norms, rather than carefully analysing the remuneration committee's detailed proposals.

The data provided in this chapter, along with the case study of AstraZeneca in Chap. 2, illustrate the complex web of agency and fiduciary relationships which exist between shareholders and directors, directors and managers, institutional investors and company boards, retail investors and investment managers, and so on. Some of these are "strong" principal-agent relationships, where an agent has been appointed by a principal under the terms of a contract that specifies the terms and conditions governing the relationship. Others are "weak" fiduciary relationships, where there is no direct contractual relationship between the two parties and the connection is more akin to that of trustee and beneficiary. The case also demonstrates that another set of agency problems arises at the funds level, where the collective action problems are even greater because holdings in retail funds are more widely dispersed and because there is far less transparency about governance and pay than in public quoted companies: the pay of investment executives is not widely publicised; some investment firms are private companies or partnerships which are not subject to the same degree of scrutiny as public corporations; in any case, the pay of executives who are not also company directors does not have to be disclosed in the detail required of public company directors. The study provides evidence of the difficulties in limiting excessive executive pay when shareholdings are widely dispersed, as predicted by the formal theory.

[20] PwC Report "ISS friend or foe to stewardship?" January 2018 https://www.pwc.co.uk/services/human-resource-services/insights/demystifying-executive-pay/iss-friend-or-foe-to-stewardship.html

α AND β FACTORS

This chapter has identified two factors that are critical in determining whether corporate governance will be successful in moderating excessive executive compensation costs. The first of these is the α factor or "principal force" which determines whether shareholders will combine together to take collective action to address agency costs. The second of these is the β factor or "fiduciary force" which determines how probable it is that non-executive directors will carry out their fiduciary responsibilities to the fullest extent possible, thus having a moderating influence on agency costs. The formal theory set out in the second section of this chapter predicts that the principal force will be at its strongest if, first, a single institutional investor's holding is sufficiently large as to make the proportionate benefits of reducing executive rents greater than the additional individual governance costs incurred in securing the reduction; secondly, an institutional investor obtains ancillary benefits from leading a collective action to recover executive rents, for example, by enhancing its reputation as an active investor; or thirdly, a group of shareholders agrees to work together to reduce executive rents and is able to spread the additional governance costs incurred in such a way that the benefits outweigh the costs in every case. The fiduciary force will be at its strongest if an individual non-executive director, for example, the chair of the remuneration committee has a powerful enough sense of their fiduciary responsibilities and professional ethics or expects to gain sufficiently valuable reputational benefits from taking a hard line on excessive compensation costs. These two factors, the principal force and the fiduciary force, are independent of each other but may operate in combination: corporate governance will be at its most effective when both α and β forces are at their strongest.

The conclusion stated here regarding the α factor is consistent with previous research on shareholder power; shareholder power has been described as a continuum extending from the relatively powerless (passive retail funds with small holdings in widely spread investment portfolios who rely, if anything, on soft activism) to the powerful (hedge funds and private equity firms who take large stakes in a small number of companies and follow a path of concentrated activism). In the middle are a number of active funds who rely on both soft activism and coordinated action.[21] It

---

[21] See, for example, the selection of essays in Hill, J., & Thomas, R. (2015). *Research Handbook on Shareholder Power*. Cheltenham, UK: Edward Elgar, in particular essays by Hill, J. (2015) and Coates, J. (2015).

is worth noting that shareholders in UK public companies possess more legal powers and participation rights than their counterparts in the US. The UK Financial Reporting Council has promoted the concept of "stewardship" by adopting a Stewardship Code in 2010 in response to a recommendation made by the Walker Review of Corporate Governance in the UK Banking Industry. The objective is to encourage investors to exercise their powers more actively, by engaging in debate with companies on their business strategies, financial performance, corporate governance and executive remuneration, as well as by voting and monitoring.[22] Nevertheless, even during the "shareholder spring" of 2012 and its mini-revival during the season of company annual general meetings in April and May 2016, UK shareholders have only succeeded in overturning executive pay proposals in a relatively small number of cases.[23]

The β factor illustrates the underlying paradox in standard agency theory of relying on the ethical motives of directors to solve agency problems. It reinforces the need, as set out in the previous chapter, to devise a more sophisticated model of economic man that recognises the significance of moral sentiments as well as economic impulses. It also gives force to the importance of developing normative models of executive and director behaviour that incorporate high deontic expectations of company directors and senior executives. By deontic, I mean expectations relating to

[22] These ideas are also consistent with proposals made in 2016 by a group of prominent public figures in the UK, led by Conservative MP Chris Philp, to establish shareholder committees, modelled on Swedish nomination committees, as part of the UK corporate governance code. They proposed that all large listed UK companies should establish committees, to be known as "shareholder committees", comprising their five largest shareholders, chaired by the largest shareholder. Shareholder committees would have three principal powers and responsibilities. Firstly, they would replace nomination committees and assume responsibility for recommending the appointment and removal of directors for a vote of all shareholders at a company's annual general meeting. This would: "make directors feel more accountable to shareholders and not to the board chairman". Secondly, they would approve the pay policy and specific pay packages proposed by the remuneration committee before they are put to a binding vote of all shareholders at AGM. This would: "allow for proper scrutiny by shareholders before the AGM vote takes place". Thirdly, shareholder committees would pose questions requiring a response by the main board, including on corporate strategy and corporate performance. This would: "formally empower shareholders to raise issues with the board, while still firmly leaving the board ultimately responsible for strategy and performance". Philp, C., (2016) "Restoring responsible ownership – Ending the ownerless corporation and controlling executive pay". *High Pay Centre*, September 2016.

[23] The Times. Boardroom pay is off the scale and shareholder revolts will not reel it back (May 4, 2016).

duty and obligations as moral concepts. In other words, we need an agency theory that focuses on the professional ethics of corporate managers and company directors, not just on material incentives.

THE CORPORATION AS COMMONS

So far in this chapter I have worked within the parameters of neoclassical economics, seeking to demonstrate, through theory and empirical analysis, that a major shortcoming of the standard model is the fact that it overlooks the collective action problem at the heart of the public corporation. Working along similar lines, Simon Deakin, Professor of Law at Cambridge, has argued in a paper entitled *The corporation as commons: rethinking property rights, governance and sustainability in the business enterprise* that the commons might provide a better foundational model for theorising about public corporations than the current combination of the standard model and legal fiction theory.[24] He draws a similar distinction to the one that I have drawn between "the firm" and "the corporation", quoting with approval the French jurist Jean-Philippe Robé, who says,

> The firm and the corporation are very often confused in the literature on the theory of the firm. The two words are often used as synonyms. They correspond, however, to totally different concepts: a corporation is a legal instrument, with a separate legal personality, which is used to legally structure the firm; a firm is an organized economic activity, corporations being used to legally structure most firms of some significance.[25]

Deakin argues that corporate law has a more central role to play in determining the nature of the corporation than the standard model envisages. He sees company law as an emergent phenomenon that has co-evolved with the emergence of corporations in industrial societies.[26] He calls for economic and legal theories of the corporation to be more empirically grounded in actual observation than the eviscerated view of legal fiction theory.[27] He concurs with my view that shareholders are not "owners" of corporations, saying,

---

[24] Deakin, S. (2012). The corporation as a commons: rethinking property rights, governance and sustainability in the business enterprise. *Queen's Law Journal*, 37 (2), pp. 339–381.

[25] Robé, J. (2011). The legal structure of the firm. *Accounting, Economics, and Law*, 1 (1), Article 5, cited by Deakin (2012), p. 352, note 31.

[26] Deakin (2012) p. 345.

[27] Deakin (2012) p. 346–347.

> Shareholders have many rights, ranging from voice and voting rights to rights in relation to distributions, which stem from the property they have in their *shares*. However, none of these rights either derives from or confers a right to property in the firm itself, or its assets, nor do any property claims which shareholders might have given them a right to manage the assets of the firm.[28]

The "agency" responsibilities of directors and executives are determined partly by company law and partly by employment law. The fiduciary responsibilities of corporate managers to the corporation, derived from common law, are more substantial than an agency perspective might imply. Other employees also have rights and responsibilities determined by employment law, and management's authority over them is conditioned by their responsibility for the physical, economic, and psychological well-being of workers.[29]

Deakin's conceptualisation of the collective action problem at the heart of the public corporation is, however, much more widely drawn than the picture I have painted in the previous section. He puts it like this:

> The firm as such cannot be owned, but in the context of the modern business enterprise, there are multiple, overlapping and often conflicting property rights or property-type claims which the legal system is meant to adjust and reconcile. As we have seen, corporate law is largely concerned with one set of such rights, those of shareholders, but this by no means exhausts the set of claims on the firm's assets. Employment law, insolvency law and fiscal law also identify claims of this kind. Each of these areas of law has a dual function: specifying the conditions under which various contributors of inputs (or, as they are sometimes called…"stakeholders") can draw on the resources of the firm while at the same time preserving and sustaining the firm's asset pool as a source of productive value. This is the sense in which the business enterprise is a "commons".[30]

In other words, in addition to shareholders, certain other persons, most notably employees (through obligations enshrined in employment law), also have rights in respect of the commons that management must respect. Corporate managers must arbitrate between these various "overlapping and conflicting" rights at the same time as they exercise their responsibility for maximising total firm value over the long term and "sustaining the

---

[28] Deakin (2012) p. 356.
[29] Deakin (2012) p. 363.
[30] Deakin (2012) p. 367–368.

firm's asset pool as a source of productive value". This is a much more holistic view of the agency responsibilities of executives.

One of the benefits of the standard model, according to its proponents, is the way that it resolves value claims between different stakeholders. Michael Jensen has argued that stakeholder theory is flawed because it violates the principle that a single value objective is a prerequisite for rational corporate strategic decision-making. He goes on to say: "a firm that adopts stakeholder theory will be handicapped in the competition for survival because, as a basis for action, stakeholder theory politicises corporations and leaves its management empowered to exercise their own preferences in spending the firm's resources."[31] The standard model tries to resolve these difficulties by allocating property rights and specifying that the primary objective of the corporation is to maximise shareholder value. This principle should be used, supporters of the standard model say, as the decision criterion for all major corporate decisions, including, for example, whether to acquiesce to a hostile takeover bid, whether to outsource a major part of a corporation's activities in the interests of cost savings, but at the expense of direct employment opportunities, whether to forgo current investment opportunities in order to benefit short-term profits, but at the expense of long-term value creation, and so on.

However, Deakin points out that there is an extensive literature describing an empirical research programme conducted over two decades, principally led by the Noble prize winner, Elinor Ostrom, which shows that the apparent contradictions and conflicts in the collective use of valuable resources can be overcome if appropriate governance and management regimes are put in place. The research on common pool resources is summarised in a collection entitled *Working Together – Collective Action, the Commons, and Multiple Methods in Practice* by Amy Poteete, Marco Janssen, and Elinor Ostrom.[32] They explain how eight design principles for the governance and management of common pool resources can be inducted from the empirical work. These design principles—summarised in Table 4.3[33]—are a rich source of ideas about the effective governance of public corporations.

[31] Jensen, M. (2001). Value maximization, stakeholder theory, and the corporate objective function. *Journal of Applied Corporate Finance,* 14 (3) p. 10.

[32] Poteete, A., Janssen, M., & Olstrom, E. (2010). *Working Together – Collective Action, the Commons, and Multiple Methods in Practice.* Princeton & Oxford: Princeton University Press, p. 100. See also Chap. 2, note 33 supra.

[33] The table is based on Olstrom, E. (2005). *Understanding Institutional Diversity.* Princeton, NJ: Princeton University Press, p. 259; Poteete et al. (2010) p. 100–101; and Deakin (2012) pp. 372 & 378.

Table 4.3 Managing the commons as a source of corporate governance design principles (after Deakin, 2012)

Design principle	Application to corporations
1. Well-defined boundaries The boundaries of the resource system and the individuals with rights to be harvest resource units should be clearly defined	The company must determine which stakeholders should have rights to participate in rule-making and value sharing, as well as what obligations it may have to people in its supply chain who are outside the formal boundaries of the firm
2. Proportionality between benefits and costs Rules specifying the amount of resource products that a user is allocated are related to local conditions and to rules requiring labour, materials, or money inputs	The principle of proportionality between inputs and benefits should apply to all significant stakeholders, not just to shareholders. This principle is particularly relevant in the event of a takeover or merger, or if special dividends are proposed
3. Collective choice arrangements Many of the individuals affected by harvesting and protection rules should be included in the group who can modify these rules	All major stakeholders should have the right to participate in rule-making and corporate governance to ensure that rules fit local contexts and are adaptable to changing circumstances
4. Monitoring Monitors who actively audit conditions and user behaviour are at least partially accountable to the users or are users themselves	Monitoring is primarily the responsibility of the board of directors, particularly non-executives. The board should recognise its obligations to stakeholders generally, not just to shareholders
5. Graduated sanctions Users who violate rules-in-use should receive graduated sanctions depending on the seriousness and context of the offence from other users or from officials accountable to these users	Sanctions for breaches of rules should be graduated and proportionate to help build trust between the board, executive management, shareholders, and other stakeholders
6. Conflict resolution mechanisms Users and their officials have rapid access to low-cost, local arenas to resolve conflict among users, or between users and officials	Corporations should build voice and conflict resolution mechanisms designed to address areas of concern or conflict quickly
7. Minimal recognition of rights The rights of users to devise their own institutions are not challenged by external governmental authorities, and users have long-term tenure rights to the resource	Shareholders and other key stakeholders should have the right to establish the governance arrangements that they regard as being most appropriate. This principle should be enabled by law and respected by governments
8. Local governance arrangements Appropriation, provision, monitoring, enforcement, conflict resolution, and governance activities should be organised in multi-layers of nested enterprises	Governance rules should reflect local circumstances, as well as state, federal, and transnational requirements. This principle is especially relevant to multinationals

The most critical of these eight design principles are discussed further below along with the implications for the governance of public corporations.

Proportionality Between Benefits and Costs

Poteete, Janssen, and Ostrom emphasise that, for the effective management of common pool resources, benefits should be allocated in proportion to inputs, for example, of capital and labour. Rules that respect proportionality are more likely to be regarded as equitable. Rules that disproportionately benefit elites will be perceived as inequitable. Perceived inequity undermines trust. Perceived fair pay is an important characteristic of high-trust organisations. If shareholders, and employees generally, perceive that senior executive pay is excessive, then their confidence in top management will be undermined. If excessive executive compensation is seen as a collective action problem, and public companies are in effect quasi-public goods, then extracting high pay for managerial elites or excessive special dividends in the short term for shareholders is like "overharvesting" in a common pool situation.

Collective Choice Arrangements

There should be broad participation in governance arrangements—individuals who are affected by resource allocation rules should have representation rights in governance systems. Voice mechanisms, such as works councils, employee advisory panels, and worker representation on company boards or major committees, can be important ways of building trust.

Monitoring

Individuals charged with monitoring should be broadly accountable, as reliable monitoring raises confidence among users of common pool resources. The board of directors must recognise its accountability to a wide range of stakeholders, including minor as well as major shareholders, employees generally, the communities in which the corporation operates, and so on.

Conflict Resolution Mechanisms

There should be rapid, local conflict resolution arrangements. Local mechanisms that allow conflicts to be aired quickly help to build trust.

Some conflicts arise simply because users interpret rules differently. Sanctions for violations of rules should be graduated. Graduated sanctions signal that infractions are notices while allowing for misunderstandings, mistakes, and exceptional circumstances. Companies should recognise that conflicts with stakeholders will inevitably arise. It is important to ensure that there are mechanisms for resolving conflicts quickly, when they do arise, and that management's mistakes are acknowledged.

Local Governance Arrangements

In much the same way that Neil Fligstein describes organisational fields as "embedded in other fields like a Russian doll",[34] so Poteete, Janssen, and Ostrom talk about "nested enterprises". They advise that, in complex common pool structures, users should be encouraged to devise their own governance arrangements as these will be best suited to local conditions. The role of governmental authorities is to enable and support local governance. This principle is relevant to the governance of multinational firms. It is consistent with ideas about self-determination and self-regulation, underpinned by the legal system, with government intervention only when it is clear that self-determination is not working effectively.

CONCLUSION

In this chapter I have explained why shareholders of public corporations are unlikely to resolve executive pay dilemmas because of collective action problems, and how ideas about the governance of common pool resources have implications for the design of effective corporate governance mechanisms. I shall return to effective corporate governance architecture in the final chapter. In the meantime, I turn in Chap. 5 to the design of executives' incentives, and to the lessons that can be drawn from behavioural science.

Further Reading

A number of the essays in Joseph Heath's book deal with collective action problems in public corporations. A good general text, which summarises

---

[34] Fligstein, N (2016) The theory of fields and its application to corporate governance. *Seattle University Law Review* 39(2) p. 242.

Mancur Olson's ideas and also covers the prisoners' dilemma, is Hardin, R. (1982|2013). *Collective Action*. Routledge. Readers may also like to refer to Olson's own seminal work, especially Parts I and II, Olson, M. (1965|1971). *The Logic of Collective Action* – Public *Goods and the* Theory *of Groups*. Harvard University Press – the situation of public corporations is addressed on pages 55–57.

REFERENCES

Becht, M., Franks, J., Mayer, C., & Rossi, S. (2008). Returns to Shareholder Activism: Evidence from a Clinical Study of the Hermes UK Focus Fund. *The Review of Financial Studies, 22*(8), 3093–3129.

Charkham, J. (1995). *Keeping Good Company: A Study of Corporate Governance in Five Countries*. Oxford: Oxford University Press.

Coates, J. (2015). Thirty Years of Evolution in the Roles of Institutional Investors in Corporate Governance. In J. Hill & R. Thomas (Eds.), *Research Handbook on Shareholder Power*. Cheltenham: Edward Elgar Publishing Limited.

Deakin, S. (2012). The Corporation as Commons: Rethinking Property Rights, Governance and Sustainability in the Business Enterprise. *Queen's Law Journal, 37*(2), 339–381.

Hardin, R. (1982|2013). *Collective Action*. London/New York: Routledge.

Heath, J. (2014). *Morality, Competition, and the Firm*. New York: Oxford University Press.

Hill, J. (2015). Images of the Shareholder – Shareholder Power and Shareholder Powerlessness. In J. Hill & R. Thomas (Eds.), *Research Handbook on Shareholder Power* (pp. 53–73). Cheltenham: Edward Elgar Publishing Limited.

Hill, J., & Thomas, R. (2015). *Research Handbook on Shareholder Power*. Cheltenham: Edward Elgar.

Jensen, M. (2001). Value Maximization, Stakeholder Theory, and the Corporate Objective Function. *Journal of Applied Corporate Finance, 14*(3), 8–22.

Levenstein, M., & Suslow, V. (2006). What Determines Cartel Success? *Journal of Economic Literature, 46*(1), 43–95.

Olson, M. (1965|1971). *The Logic of Collective Action – Public Goods and the Theory of Groups*. Cambridge, MA: Harvard University Press.

Poteete, A., Janssen, M., & Olstrom, E. (2010). *Working Together – Collective Action, the Commons, and Multiple Methods in Practice*. Princeton/Oxford: Princeton University Press.

Robe, J. (2011). The Legal Structure of the Firm. *Accounting, Economics, and Law, 1*(1), 1–86, Article 5.

CHAPTER 5

Behavioural Agency Theory

Abstract This chapter explains that the conventional design of executive compensation plans, involving high salaries, generous bonuses, and highly leveraged stock programmes is based on an outdated set of assumptions about human behaviour and executive agency. It describes a revised theory of agency and a modified design framework for executive pay plans based on developments in behavioural science.

Keywords Long-term incentive plans • Risk aversion • Time discounting • Behavioural agency theory

INTRODUCTION

It will now be apparent that a central thesis of this book is that the conventional design of executive compensation plans, involving high salaries, generous bonuses, and highly leveraged stock programmes is based on an outdated model of executive agency. The standard theory assumes that executives are rational, self-interested, utility maximisers, motivated only

---

This chapter draws largely from Pepper, A. (2017). Applying economic psychology to the problem of executive compensation. *The Psychologist-Manager Journal*, 20(4), 195–207. The relevant parts are reprinted with the permission of the American Psychological Association.

© The Author(s) 2019
A. Pepper, *Agency Theory and Executive Pay*,
https://doi.org/10.1007/978-3-319-99969-2_5

by money. It postulates that companies must provide high-powered, performance-based incentives in order to align the interests of shareholder and executives.[1] Yet, as has already been explained, we have known for some time that agency theory has major shortcomings.[2] The data indicate that executive compensation is correlated with firm size, not company profits. Conventional wisdom today is that CEO pay increases as a power function of company size. Some economists argue that this conclusion, that *ex post* executive pay is correlated with firm size, is still consistent with optimal contracting even if, *ex ante*, firms are trying to link executive pay to firm performance, but this is like arguing that a man travelling from London to Edinburgh who takes a wrong turning and arrives in Glasgow instead has achieved an efficient outcome because he has still made it to Scotland.

This chapter proposes a new version of agency theory that provides a better explanation of the connection between executive compensation, agent performance, firm performance, and the interests of shareholders. This is the "behavioural agency model" or "behavioural agency theory", which develops a line of argument first advanced by Robert Wiseman and Luis Gomez-Mejia in 1998.[3] They proposed that the normal risk assumptions of agency theory should be varied to incorporate ideas about loss aversion. Sanders and Carpenter adopted a behavioural agency theory perspective in their examination of stock repurchase programme announcements.[4]

---

[1] The extensive literature on the application of agency theory to executive compensation dates back to Jensen, M., & Meckling, W. (1976). Theory of the firm: Managerial behavior, agency costs and ownership structure. *Journal of Financial Economics, 3*(4). A helpful summary is provided by Eisenhardt, K. (1989) Agency Theory: An Assessment and Review, *Academy of Management Review,* 14 (1), 57–74.

[2] In 1990, in an article entitled, "Performance pay and top-management incentives", published in the *Journal of Political Economy*, Michael Jensen and Kevin Murphy were unable to find a statistically significant connection between CEO pay and performance. Ten years later Henry Tosi, Steven Werner, Jeffrey Katz, and Luis Gomez-Mejia, in "How much does performance matter? A meta-analysis of CEO pay studies", in the *Journal of Management,* concluded that incentive alignment as an explanatory agency construct for CEO pay was at best weakly supported by the evidence based on their meta-analysis of over 100 empirical studies. In 2010 a literature review by Carola Frydman and Dirk Jenter entitled, "CEO compensation", in the *Annual Review of Financial Economics,* concluded that neither agency theory nor the alternative "managerial power hypothesis" proposed by Lucien Bebchuk, Jesse Fried, and David Walker (2002) was fully consistent with the available evidence.

[3] Wiseman, R., & Gomez-Mejia, L. (1998). A behavioral agency model of managerial risk taking. *Academy of Management Review,* 23 (1), 133–153.

[4] Sanders, G., & Carpenter, M. (2003). A behavioral agency theory perspective on stock repurchase program announcements. *Academy of Management Journal,* 46 (3), 160–178.

Rebitzer and Taylor provide a general examination of behavioural approaches to agency and labour markets in the fourth edition of Ashenfelter and Card's influential handbook on labour economics, published in 2011.[5] In addition, I published various papers on behavioural agency theory between 2013 and 2017 with Julie Gore and a number of other collaborators.[6]

In contrast to the standard agency framework, which focuses on monitoring costs and incentive alignment, behavioural agency theory places agent performance and work motivation at the centre of the agency model, arguing that the interests of shareholders and their agents are most likely to be aligned if executives are motivated to perform to the best of their abilities, given the available opportunities. It builds on four modifications to the standard theory of agency that have been identified as key factors affecting behaviour by behavioural economists and economic psychologists. These modifications relate to

- Risk and uncertainty;
- Temporal discounting;
- Fairness and inequity aversion;
- Intrinsic and extrinsic motivation.

Goal-setting, regarding by psychologists as an important part of the theory of motivation, has also been linked with the behavioural agency model, on the grounds that it represents a pragmatic way of contracting between principal and agent.[7]

[5] Rebitzer, J., & Taylor, L. (2011). Extrinsic rewards and intrinsic motives: standard and behavioral approaches to agency and labor markets. In O. Ashenfelter & D. Card (Eds.), *Handbook of Labor Economics* (Vol. 4A, pp. 701–772). Amsterdam: North-Holland.

[6] Pepper, A., & Gore, J. (2014). The economic psychology of incentives – an international study of top managers. *Journal of World Business, 49*(3), 289–464; Pepper, A., & Gore, J. (2015). Behavioral agency theory: New foundations for theorizing about executive compensation. *Journal of Management, 41*(4), 1045–1068; Pepper, A., Gore, J., & Crossman, A. (2013). Are long-term incentive plans an effective and efficient way of motivating senior executives? *Human Resource Management Journal, 23*(1), 36–51; Pepper, A., Gosling, T., & Gore, J. (2015). Fairness, envy, guilt and greed: building equity considerations into agency theory. *Human Relations, 68*(8), 1291–1314. Pepper, A. (2015). *The Economic Psychology of Incentives – New Design Principles for Executive Pay*. Basingstoke, UK: Palgrave Macmillan. Pepper, A. (2017). Applying economic psychology to the problem of executive compensation. *The Psychologist-Manager, 20*(4), 195–207.

[7] Pepper & Gore (2015). For goal setting theory, see Locke, E., & Latham, G. (1990). *A theory of goal setting and task performance*. Englewood Cliffs, NJ: Prentice Hall.

MODIFYING THE ASSUMPTIONS OF AGENCY THEORY

Empirical research associated with behavioural agency theory has provided a better understanding of the relationship between executives' pay and their motivation.[8] A number of key points have emerged from the research. First, executives are much more risk averse than financial theory predicts, preferring fixed outcomes to risky, yet potentially more rewarding alternatives. They also attach a heavy discount to ambiguous and complex incentives. Secondly, executives are very high time discounters, typically marking down the value of complex long-term incentives at a rate in excess of 30% per year. Thirdly, fairness matters—executives are as concerned about the perceived equity or inequity of their awards relative to peers as they are with absolute amounts. Finally, intrinsic motivation is much more important than admitted by traditional economic theory, to the point where many executives would give up around 28% of their income to work in more satisfying roles.

Whereas agency theory focuses on how incentive contracts can be best designed to align the interests of shareholders (principals) and executives (agents), behavioural agency theory focuses on agent motivation. The theory of work motivation most commonly used by psychologists when investigating the motivational impact of monetary incentives is expectancy theory, originally advanced in the 1960s by the American psychologist Victor Vroom. A modern version of expectancy theory is temporal motivation theory, devised by management scholars Piers Steel and Cornelius König. This incorporates George Ainslie's theory about hyperbolic time discounting and Daniel Kahneman and Amos Tversky's prospect theory.[9] Temporal motivation theory postulates that a person's motivation to carry out a particular act is the product of his or her expectancy that the act will lead, directly or indirectly, to a particular outcome, and the value which he or she attaches to that outcome, discounted for risk, and for any time delay between the occurrence of the final outcome and the initial act.

---

[8] See in particular Pepper and Gore (2014); Pepper et al. (2013); Pepper et al. (2015).

[9] Voom, V., (1964) *Work and Motivation*. New York: Wiley; Steel.P., & Konig, C., (2006) "Integrating Theories of Motivation", *Academy of Management Review*, 31 (4): 889–913; Ainslie, G., (1991) "Derivation of Rational Economic Behavior from Hyperbolic Curves", *American Economic Review*, 81 (2): 334–340; Kahneman, D., & Tversky, A., (1979) "Prospect Theory – An Analysis of Decision Under Risk", *Econometrica*, 47 (2): 263–29.

Risk and Uncertainty

In one study which I carried out with my collaborator Julie Gore, participants were asked if they would rather have (A) a 50% chance of receiving a bonus of $90,000; otherwise nothing; or (B) $41,250 for certain? The expected value of (A) is $45,000, suggesting that a risk-neutral executive should prefer (A) over (B). Yet in our research 63% of executives chose (B), representing a risk premium of around 9%. By asking similar questions, it was possible to demonstrate that executives required a risk premium of up to 17% before selecting the risky option. To put this in context, rational choice risk premiums have been estimated at between 6% and 11% for executives with up to 50% of their wealth tied up in firm equity—the risk premiums implied by the questions in the research were therefore at or above the upper end of this range.[10] Executives also attach a heavy discount to uncertain, ambiguous, and complex incentives. One executive said of performance-based stock programmes: "because of complexity, direct motivation is often not there on a day-to-day basis". You cannot be extrinsically motivated by something which you do not understand.

Temporal Discounting

According to the standard theory, individuals should discount future receipts at rates that are consistent with the return on comparably risky future cash flows, adjusted for inflation. At the time empirical work was carried out temporal discount rates should have been close to the risk-free rate of around 1% per annum. However, there was evidence in the study that executives discounted for time at much higher rates, with a median of 33%. This is consistent with the thesis that psychologically we discount for the future hyperbolically rather than exponentially. As another executive put it: "long-term incentives are an amount of money with a very high discount attached to it".

Figure 5.1 shows both hyperbolic (perceived value) and exponential (economic value) utility functions, along with the gap between the economic value and perceived value of long-term incentives, which only closes when the final pay-out occurs. Value is created in product markets when

[10] Conyon, M., Core, J., & Guay, G., (2011) "Are U.S. CEOs Paid More Than U.K. CEOs? Inferences from Risk-adjusted Pay". *The Review of Financial Studies*, 24 (2): 402–438.

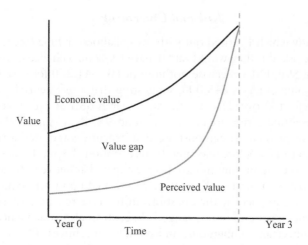

Fig. 5.1 Exponential and hyperbolic discounting

the amount a customer is willing to pay for a product or service is greater than the cost of providing that product or service, the surplus being shared between the supplier (profit) and the buyer (the customer's surplus). In a similar way, to the extent that a principal remunerates an agent such that the value of an award perceived by the agent is greater than the cost to the principal, then value is created; conversely, if a principal remunerates an agent in such a way that the cost to the principal is greater than the value perceived by the agent, then the value is destroyed. The gap between the two curves in Fig. 5.1 represents value destruction. As another executive put it, "we are being rewarded in a currency we don't value."

Intrinsic Motivation

Questions about the relationship between intrinsic and extrinsic motivation have provoked a range of responses. Another study found that, for senior executives, certain intrinsic factors, especially an orientation towards achievement, are important primary sources of behaviour. Power-status and intimacy-teamwork were also mentioned by executives as significant factors affecting the way people behave. However, intrinsic needs or drives should not be seen as substitutes for extrinsic rewards: a substantially minimum level of remuneration must be provided. One CEO described it like this: "Once you are at a threshold level on the financial structures, a level which is felt to be fair and appropriate to the market, then [intrinsic factors] become really important...but if you are at a significant discount on the monetary part then the other things will not make up for it". Nevertheless, a number of executives commented that very large awards should not be necessary to engage and motivate executives. A company chairman, commenting on the US executive labour market, said: "I do not believe, nor have I ever observed, that $100 million motivates people more than $10 million, indeed more than $1 million". In practice, the relationship between intrinsic and extrinsic motivation is complex and hard to unravel. As well as providing material benefits, extrinsic rewards are also important sources of information for executives. They give signals which executives can use to measure their value relative to their peers, how highly they are valued by their company boards, and even in some cases their self-worth. As another executive put it: "the principal role of money is...as a way of keeping the score".

Some scholars argue that intrinsic and extrinsic motivation are neither independent nor additive, proposing instead that contingent monetary rewards might actually cause a reduction in intrinsic motivation. Jeffrey Pfeffer, an American business theorist, contends that large external rewards can actually undermine intrinsic motivation.[11] Similarly, Bruno Frey, a Swiss behavioural economist, postulates that extrinsic rewards may "crowd-out" intrinsic motivation: people become distracted by monetary rewards, particularly if incentives are badly designed.[12] As one executive said: "if the amounts are large enough they can make one lose sight of the intrinsic". Empirical data shows that on average executives would be prepared to sacrifice around 28% of their earnings if they worked in a more ideal job.

[11] Pfeffer, J., (1998) "Six Dangerous Myths About Pay", *Harvard Business Review*, 76: 106–120.

[12] Frey, B., (1997) *Not Just for Money, an Economic Theory of Personal Motivation*, Cheltenham, UK: Edward Elgar Publishing; Frey, B., & Jegen, R., (2001) "Motivation Crowding Theory". *Journal of Economic Surveys*, 15 (5): 589–611.

Fairness and Inequity Aversion

Scholarly work in a number of academic traditions has demonstrated that fairness is a key factor in determining whether employees are satisfied with their pay, especially when comparisons are made with the compensation of other team members.[13] Yet fairness as between senior executives, especially among top-management teams, has not generally featured in theoretical accounts of executive incentives; equity considerations play no part in standard agency theory.

An important way in which rewards are evaluated is by drawing comparisons with salient others. In one of our studies, executives commented as follows: "internal relativity is a big issue"; "the only way I really think about compensation is 'do I feel fairly compensated relative to my peers?'" and "corporate executives appear to be very sensitive to differentials with perceived peers". Agency theory should pay more attention to fairness and social comparisons.[14]

Goal-Setting, Contracting, and Monitoring

Goal-setting, contracting, and monitoring are also integral to the principal-agent relationship: goal-setting and monitoring are important factors in legal contracting, which is a key element in the link between principal and agent; they have also been demonstrated to be an important component of agent motivation. Goal-setting theory postulates a strong connection between goals, commitment, and performance. Goals must be specific, difficult, attainable, and self-set or explicitly agreed to for the motivational affect to be maximised. Much of the empirical work supporting goal-setting theory has been carried out in an industrial context (e.g.,

[13] The most famous article about the impact of fairness motivation in the context of compensation is by John StaceyAdams entitled, "Inequity in social exchange", found in L. Berkowitz (Ed.), (1965) *Advances in experimental social psychology.* Academic Press, New York. Other helpful references, in various academic traditions, include Festinger, L., (1954) "A Theory of Social Comparison Processes," *Human Relations,* 7(2):117–140; Varian H., (1974) "Equity, Envy and Efficiency". *Journal of Economic Theory,* 9(1): 63–91, and (1975) "Distributive Justice, Welfare Economics and the Theory of Fairness. *Philosophy and Public Affairs,* 4(3): 223–247; Fehr, E., & Schmidt, K., (1999) "A Theory of Fairness, Competition, and Cooperation". *The Quarterly Journal of Economics,* 114(3): 817–868; Folger, R., & Cropanzano, R., (2001) "Fairness Theory: Justice as Accountability", in J. Greenberg & R. Cropanzano (eds.), *Advances in Organization Justice,* Stanford, CA: Stanford University Press.
[14] For more details of the research on fairness, see Pepper et al. 2015.

with loggers, truck drivers, and word processing operators). Nevertheless, many of the features of goal-setting theory are generalisable to senior executives. Edwin Locke and Gary Latham, two famous goal-setting theorists, make three points which are particularly pertinent to agency relationships.[15] First, they argue that monetary incentives enhance goal commitment, but have no substantive effect on motivation unless linked to goal-setting and achievement. Secondly, they explain, through a model that they call the "high performance cycle", how goal-setting and achievement together lead to high performance, in turn leading to rewards, high job-satisfaction, and self-efficacy. Thirdly, they suggest a possible connection with Kahneman and Tversky's prospect theory, as both theories stress the importance of reference points in cognition.

One of the main problems with the relationship between principals and agents which has been identified by agency theorists is that agency contracts are inevitably incomplete.[16] If principals were able to specify completely all that they required of their agents, then there might be no need for incentive contracts to align the interests of principals and agents—monitoring of actions and outcomes might suffice. However, in practice, there are limits on knowledge and cognition. One of the reasons that principals employ agents is for the agents' expertise. An agent who is more knowledgeable about the matters that are to be specified in a contract may be able to second-guess the principal during and after contract negotiation to the agent's advantage and the principal's detriment. There are also dynamic constraints. Over the course of time, the business environment which provides the backdrop for the agency contract inevitably changes. Actions that are contractually required of the agent when a contract is negotiated may cease to be appropriate at a later date because of environmental changes, and other actions which could not have been anticipated ex ante may subsequently become necessary ex post. It is contractual uncertainties of this kind that the economist John Roberts is referring to when he advocates the merits of weak rather than strong incentives in

---

[15] Locke, E., & Latham, G. (2002). Building a practically useful theory of goal setting and task motivation – a 35 year odyssey. *American Psychologist, 57,* 705–717.

[16] Grossman, S., & Hart, O. (1983). An analysis of the principal-agent problem. *Econometrica,* 51(1), 7–45; Grossman & Hart (1983) Hart, O. (1995). *Firms, contracts and financial structure.* Oxford University Press.

agency relationships.[17] Goal-setting, especially when it involves discussions between principal and agent about the appropriate level of objectives, is a pragmatic way of contracting, given limits on knowledge and cognition. It is also a signalling mechanism, indicating to one of the parties in an exchange relationship, the agent, what is required by another party, the principal. Michael Spence has shown how signalling mechanisms of this kind form an important part of an economic exchange in the context of employment.[18] Thus, goal-setting, monitoring, and reward, as part of a regular high-performance management cycle, provide a way of improving the quality of contracting in a manner which helps to enhance rather than undermine agent motivation.

Agents' Job Performance and the Work Motivation Cycle

The various elements of the sub-system which models agent job performance and work motivation are summarised in Fig. 5.2. It illustrates the trade-off between intrinsic and extrinsic motivation, the roles played by risk, time discounting, and inequity aversion. The goal-setting, contracting and monitoring processes are illustrated, along with an integral feedback mechanism.

Figure 5.2 puts agent motivation rather than incentive alignment at the heart of behavioural agency theory. It underlines the importance of encouraging, and not undermining, positive agent behaviours. Agent's job performance in turn contributes to the firm's performance, given the important roles which senior executives play. Figure 5.3 graphs the relationship between pay and motivation according to behavioural agency theory. It illustrates how total motivation is the sum of the intrinsic and extrinsic motivation curves. It shows the incentive 'sweet spot' (A), where the motivational benefit of an additional dollar of pay is maximised, as well as point (B) when 'crowding out' sets in, after which intrinsic motivation is undermined by each additional dollar of incentive pay, and total motivation therefore declines.

[17] Roberts, J. (2010). Designing incentives in organizations. *Journal of Institutional Economics,* 6 (1), 125–132.

[18] Spence, M. (1973). Job market signalling. *Quarterly Journal of Economics,* 87 (3) pp. 355–374.

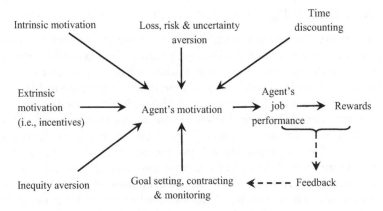

Fig. 5.2 Agent's job performance and work motivation cycle

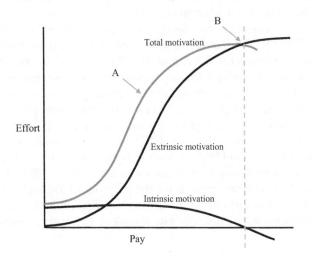

Fig. 5.3 Agent's pay-effort curve

New Design Principles for Executive Compensation

Agency theory has focused much attention on the use of high-powered incentives as a mechanism for overcoming agency costs in public corporations. In so doing, economists and finance scholars have dramatically underplayed the role of psychology in determining organisational behaviour. Considerable time has been spent devising highly elaborate incentive plans, which the philosopher Joseph Heath, in an article entitled, "The uses and abuses of agency theory", describes as being of "baroque complexity",[19] while neglecting risk perceptions, time discounting, and intrinsic motivation. Inflation in executive pay over the last 30 years is almost entirely related to pay-outs from stock options and other long-term incentive plans: senior executives' salaries have been remarkably stable for many years. Pepper and Gore's research suggests, on the face of it somewhat perversely, that companies would be better paying *larger* salaries, using annual cash bonuses to incentivise desired actions and behaviours, and avoiding performance-related equity plans altogether. Executives should be required to invest their bonuses in company shares until they have sufficient "skin in the game" to align their interests with shareholders. Alternatively, for greater tax efficiency, annual bonuses might be provided in the form of restricted stock, with time constraints on vesting but without financial performance metrics or restrictions on receiving dividends, until holding requirements have been met. The critical point here is that holders of restricted shares must feel like "owners" in order to avoid simply replicating the time discounting problem that exists with long-term incentives. Restricted shares should therefore ideally be given dividend and representation rights from the start but have constraints placed on the right to sell shares during a (relatively long) vesting period.

To illustrate, imagine that the CEO in a large company currently receives a salary of £500,000, an annual bonus opportunity of 200% of salary, and an annual long-term incentive plan award of 400% of salary.[20] Pensions and benefits are ignored for the purposes of simplicity. The face value of the compensation package is therefore £3,500,000. Assume that the CEO has a subjective discount rate for risk of 16% and for time of 33%.

---

[19] Heath, J., (2014) "The Uses and Abuses of Agency Theory", in *Morality, Competition, and the Firm – The Market Failures Approach to Business Ethics,* New York: Oxford University Press.

[20] The calculations in this section are based on Table 1 in Pepper (2017).

After these discounts have been applied the subjective value of the bonus is reduced to £562,500. The perceived value of the long-term incentive, discounted over three years at a rate of 33% per annum, as well as for risk, is reduced to £500,000. Thus, the total subjective value of the CEO's current compensation package amounts to around £1,562,500. The accounting cost to the company, assuming the bonus and long-term incentive both pay out at a rate of 75% and that the fair value of the long-term incentive at the date of grant is broadly the same as the amount which is eventually disbursed, is around £2,750,000.

By redesigning the compensation pack according to the new design principles set out in this chapter, the same subjective value of £1,562,500 can be delivered to the executive at a lower total cost to the company and with a lower headline rate of executive pay. The redesigned package comprises a base salary of £1,000,000, twice the amount payable under the traditional arrangements, and an annual bonus opportunity of 100% of salary. By the time the value of the bonus has been discounted for risk by 16% and for time by 33%, its perceived value is again reduced to around £562,500. Assuming that the actual bonus pays out at a rate of 75%, both the cost to the company and headline rate of executive compensation is reduced to £1,750,000.

One of the main objectives of incentive contracts under agency theory is to align the interests of shareholders and managers in order to reduce agency costs. Alignment of the CEO's interests with those of the corporation's shareholders is typically obtained by requiring the CEO to invest some of his available after-tax cash in company shares until a meaningful shareholding has been obtained, combined with participation in a long-term incentive plan. In the example set out above, under a traditional compensation package, on the basis that the executive is required to buy shares with a value at the date of acquisition equivalent to 200% of salary, and assuming a tax rate of 40%, the combination of shares and LTIPs represents around three years of free cash flow. The shareholding, combined with exposure under the long-term incentive plan, means that at any one time the CEO will have an interest in around £3,000,000 worth of shares in the company. Under the new design principles, a similar level of exposure to own-company shares can be obtained by investing after-tax free cash flow over a period of around four years. To ensure continuing alignment of interests, it should be a requirement that the shareholding is retained throughout the executive's term of office and for a period of one or two years after employment has ceased.

At least one major institutional investor in the UK has recognised the merits of this approach. The Norwegian sovereign wealth fund has published guidelines for the remuneration of CEOs of the companies in which it invests which are consistent with, and in part based upon, the research described in this chapter.[21] A company whose executive reward strategy is consistent with many of the design principles described here is Berkshire Hathaway. In a number of his famous letters to shareholders, Warren Buffett has explained how Berkshire Hathaway has adopted an incentive compensation system which rewards key managers with generous salaries and cash bonuses, but which eschews equity plans. At Berkshire, salaries are calibrated according to the size of the executive's job, and cash bonuses are paid annually for meeting targets within the executive's own business unit. Performance is defined in different ways depending on the economics of the underlying business, but Buffett says he tries to keep things "simple and fair". Business unit performance is rewarded whether Berkshire stock rises, falls, or stays the same. Managers are encouraged to buy Berkshire stock with their bonuses, and Buffett notes that many have done so, thus benefitting from the strong sustained share price performance of Berkshire Hathaway over many years. By buying stock with their own money, managers accept the risks and carrying costs of ownership as well as benefitting from dividends and opportunities for capital growth. In this way their interests are much more closely aligned with those of other shareholders than would be the case if they were beneficiaries of stock option awards or other types of equity incentive.[22]

Executive compensation has become a major political issue and many believe that reform is vital to restore faith in capitalism. Businesses are waking up to the fact that long-term incentive plans do not work as intended. How many non-executives on board compensation committees really understand the formula they are approving and the size of the awards that may crystallise in future as a result? According to Philip Hampton, Chairman of GlaxoSmithKline plc., "we've probably been going in the

[21] See Norges Bank Investment Management, Asset Manager Perspective 01|2017, "Remuneration of the CEO", published April 7, 2017.
[22] See Buffett, W., (2014) *Berkshire Hathaway – Letters to Shareholders 1965–2013*, Palo Alto, CA: Max Olson; and Buffett, W., (2014) *The Essays of Warren Buffett – Lessons for Investors and Managers*, Singapore, John Wiley & Sons.

wrong direction for 20 years or more".[23] Change is self-evidently necessary. By incorporating the design principles set out in this chapter into their thinking about executive compensation, companies might be encouraged to move towards what would in aggregate be smaller, but more balanced, more effective compensation plans, benefitting business and society as a whole, yet without fundamentally undermining the motivation of our top executives.

This chapter has addressed incentive contracts, one of the solutions to the agency problem proposed by Jensen and Meckling. In the final chapter I turn to the other standard solution to the agency problem, monitoring by principals of the activities of agents (or corporate governance as we now know it), and suggest ways in which this too might be made more effective.

Further Reading
The research behind this chapter is described in more detail in: Pepper, A. (2015). *The Economic Psychology of Incentives – New Design Principles for Executive Pay*. Palgrave Macmillan.

REFERENCES

Buffett, W. (2014a). *Berkshire Hathaway – Letters to Shareholders 1965–2013* (M. Olson, Ed.). Palo Alto: Max Olson.

Buffett, W. (2014b). *The Essays of Warren Buffett – Lessons for Investors and Managers* (L. Cunningham, Ed.). Singapore: John Wiley & Sons Singapore Pte. Ltd.

Eisenhardt, K. M. (1989). Agency Theory: An Assessment and Review. *Academy of Management Review, 14*(1), 57–74.

Frey, B. (1997). *Not Just for Money, an Economic Theory of Personal Motivation*. Cheltenham: Edward Elgar Publishing.

Frey, B., & Jegen, R. (2001). Motivation Crowding Theory. *Journal of Economic Surveys, 15*, 589–611.

Grossman, S., & Hart, O. (1983). An Analysis of the Principal-Agent Problem. *Econometrica, 51*(1), 7–45.

---

[23] Philip Hampton was quoted in the Financial Times on May 9, 2016 by Financial Editor Patrick Jenkins.

Hart, O. (1995). *Firms, Contracts and Financial Structure.* Oxford: Oxford University Press.

Locke, E., & Latham, G. (1990). *A Theory of Goal Setting and Task Performance.* Englewood Cliffs: Prentice Hall.

Locke, E., & Latham, G. (2002). Building a Practically Useful Theory of Goal Setting and Task Motivation – A 35 Year Odyssey. *American Psychologist, 57,* 705–717.

Pepper, A. (2015). *The Economic Psychology of Incentives – New Design Principles for Executive Pay.* Basingstoke: Palgrave Macmillan.

Pepper, A. (2017). Applying Economic Psychology to the Problem of Executive Compensation. *The Psychologist-Manager, 20*(4), 195–207.

Pepper, A., & Gore, J. (2014). The Economic Psychology of Incentives – An International Study of Top Managers. *Journal of World Business, 49*(3), 289–464.

Pepper, A., & Gore, J. (2015). Behavioral Agency Theory: New Foundations for Theorizing About Executive Compensation. *Journal of Management, 41*(4), 1045–1068.

Pepper, A., Gore, J., & Crossman, A. (2013). Are Long-Term Incentive Plans an Effective and Efficient Way of Motivating Senior Executives? *Human Resource Management Journal, 23*(1), 36–51.

Pepper, A., Gosling, T., & Gore, J. (2015). Fairness, Envy, Guilt and Greed: Building Equity Considerations into Agency Theory. *Human Relations, 68*(8), 1291–1314.

Rebitzer, J., & Taylor, L. (2011). Extrinsic Rewards and Intrinsic Motives: Standard and Behavioral Approaches to Agency and Labor Markets. In O. Ashenfelter & D. Card (Eds.), *Handbook of Labor Economics* (Vol. 4A, pp. 701–772). Amsterdam: North-Holland.

Roberts, J. (2010). Designing Incentives in Organizations. *Journal of Institutional Economics, 6*(1), 125–132.

Sanders, G., & Carpenter, M. (2003). A Behavioral Agency Theory Perspective on Stock Repurchase Program Announcements. *Academy of Management Journal, 46*(3), 160–178.

Spence, M. (1973). Job Market Signalling. *Quarterly Journal of Economics, 87,* 355–374.

Wiseman, R., & Gomez-Mejia, L. (1998). A Behavioral Agency Model of Managerial Risk Taking. *Academy of Management Review, 23,* 133–153.

CHAPTER 6

The Modern Corporation's Final Chapter

Abstract The final section concludes by drawing together the various ideas about corporate governance and incentives that have been identified earlier in the book, and shows how these ideas are consistent with proposals for a possible future for the public corporation set out in the last chapter of *The Modern Corporation and Private Property*, by Berle and Means, published in 1932.

Keywords The modern corporation • Rethinking capitalism

INTRODUCTION

At the end of *The Modern Corporation and Private Property*, Adolf Berle and Gardiner Means describe three possible futures for the public corporation. First, they propose that the traditional logic of property rights, whereby corporations "belong" to their shareholders, might be substantially reinforced, such that managers controlling corporations are placed explicitly in the position of trustees who are required to operate the corporation for the sole benefit of shareholders. This would require corporate law and securities regulation to be tightened considerably to enshrine in law a doctrine which Berle and Means refer to as "corporate powers as powers in trust":

A. Pepper, *Agency Theory and Executive Pay*,
https://doi.org/10.1007/978-3-319-99969-2_6

By application of this doctrine, the group in control of a corporation would be placed in a position of trusteeship in which it would be called on to operate or arrange for the operation of the corporation for the sole benefit of the security owners despite the fact that the latter have ceased to power over or to accept responsibility for the active property in which they have an interest. Were this course followed, the bulk of American industry might soon be operated by trustees for the sole benefit of inactive and irresponsible security owners.[1]

It is clear from the way that this paragraph concludes (the reference to "inactive and irresponsible security owners") that Berle and Means do not favour this first option. "Inactive and irresponsible" shareholders do not deserve the benefit of full fiduciary oversight.

Berle and Means like the second option even less.[2] They describe how the inexorable logic of laissez-faire economics and pursuit of the profit motive might lead to "drastic conclusions":

If, by reason of these new relationships, the men in control of a corporation can operate it in their own interests, and can divert a portion of the asset fund or income stream to their own uses, such is their privilege. Under this view, since the new powers have been acquired on a quasi-contractual basis, the security holders have agreed in advance to any losses which they may suffer by reason of such use.[3]

To put this in another way, if shareholders' reasonable expectations are satisfied by receiving regular dividends and having the ability to sell securities at any time on the stock market, then the rent-seeking activities of managers should be regarded as an inevitable and acceptable cost of investing in company shares. Under this scenario, investors would simply have to live with "rentier capitalism".

To many people, this second possible future, characterised as it is by powerful rent-seeking managers, describes rather well the current state of Western capitalism. It is not, to the liberal-minded and socially conscious,

---

[1] Berle, A., & Means, G. (1932) *The Modern Corporation and Private Property*. New York: Macmillan. p. 354.

[2] As they say, "if these were the only alternatives, the former would appear to be the lesser of two evils". Berle, A., & Means, G. (1932) p. 355.

[3] Berle, A., & Means, G. (1932) p. 354.

an attractive option. While many would regard capitalism as having been the most successful wealth-creating system that the world has ever seen, its current version appears to have a number of undesirable features. There are also signs of stress. In June 2018 the Bagehot column in the Economist newspaper put it like this: "wage growth is sluggish; economic insecurity is rife; a well-connected oligarchy is sucking up a disproportionate share of the proceeds of growth".[4] Bagehot goes on to describe how, in *The Wealth of Nations*, Adam Smith, the father of modern economics, worried that markets were (in Bagehot's words) "prone to being hijacked by rent-seekers". These potential rent-seekers, according to Smith, may include senior executives of companies with dispersed shareholdings, where "negligence and profusion"[5] prevails. Some would argue that this is indeed what has happened at the start of the 21st century.

An extensive literature on the present state of capitalism has grown since the financial crisis of 2008–2009. Certain commentators, including Guy Standing, Wolfgang Streeck, and Paul Mason, have predicted the end of capitalism as we know it.[6] Others, including Martin Wolf, Michael Jacobs, and Mariana Mazzucato, have provided a more nuanced analysis—they argue that predictions of capitalism's imminent demise are greatly exaggerated, while at the same time acknowledging that some fundamental changes are required to the current economic system in the West.[7] The economist and social commentator John Kay has been saying much the same thing for some time.[8] In a similar spirit, Jesse Norman argues that Adam Smith was not the market fundamentalist and apologist for inequality and human selfishness that some neoliberal economists claim that he

[4] The Economist. Good capitalism v bad capitalism (June 9, 2018) p. 30.

[5] Adam Smith (1776) *An Inquiry into the Nature and Causes of the Wealth of Nations.* Book V, Chapter 1, Part III.

[6] Standing, G. (2016). *The Corruption of Capitalism: Why Rentiers Thrive and Work Does Not Pay.* London: Biteback Publishing Limited., Streeck, W. (2016). *How Will Capitalism End? Essays on a Failing System.* London: Verso., Mason, P. (2015). *Post Capitalism: A Guide to our Future.* London: Allen Lane.

[7] Wolf, M. (2014). *The Shifts and the Shocks: What We've Learned – And Have Still To Learn – From the Financial Crisis.* London: Allen Lane. Jacobs, M., & Mazzucato, M. (2016). *Rethinking Capitalism.* Chichester: Wiley Blackwell. Mazzucato, M. (2018). *The Value of Everything: Making and Taking in the Global Economy.* London: Allen Lane.

[8] See, for example, Kay, J. (2003). *The Truth About Markets: Their Genius, Their Limits, Their Follies.* London: Allen Lane.

is.⁹ Smith's second great work, *The Theory of Moral Sentiments*, anticipates a number of ideas subsequently found in modern behavioural economics.¹⁰ Far from being a doctrinaire libertarian, Smith would, according to Norman, have supported many of the proposals for repairing capitalism advanced by Wolf, Jacobs, Mazzucato, and Kay.

Karl Marx famously predicted the end of capitalism in the nineteenth century. He was wrong of course. The weight of evidence does not support communism, the alternative to capitalism that was proposed by Marx: the collapse of the Soviet Union in 1990 brought to an end its communist command economy, sometimes described as the greatest (failed) field experiment of twentieth-century economics. Central planning on such a grand scale does not work. Capitalism is fixable but changes are required. The thesis of this book is that one of the areas where change is necessary is in corporate governance and executive compensation.

The Aspirin Trap

By proposing ever-larger awards to incentivise senior executives, especially under long-term incentive plans, agency theorists have fallen into the "aspirin trap". Let me explain what I mean by this. One 300 mg aspirin tablet will cure your headache. Two or three will do so more quickly. Taking 20 tablets at one time will make you ill. A single dose of 50 tablets might kill you. This phenomenon, involving a favourable response to a low level of exposure of a potentially toxic substance but negative responses to much larger exposures, is called "hormesis" by biologists. It exemplifies how relationships in nature are rarely defined by linear functions.

The human motivation curve is not a linear function, as I have explained in Chap. 5. When it comes to pay, relatively small (proportionately speaking) extrinsic incentives can help to enhance agent motivation because they signal what is most valued by principals. Larger incentives may increase this motivational effect by increasing the strength of the signal and providing a tangible reward. However, at some point, extrinsic incentives start to undermine intrinsic motivation, and eventually intrinsic motivation may be crowded out altogether. Very large incentive payments can have undesirable

⁹ Norman, J. (2018). *Adam Smith: What He Thought and Why it Matters*. London: Allen Lane.

¹⁰ Ashraf, N., Camerer, C., & Loewenstein, G. (2005). Adam Smith, Behavioral Economist. *Journal of Economic Perspectives, 19*(3), 131–145.

consequences. The size and nature of awards recommended by agency theorists in order to encourage high performance and align the interests of shareholders and managers under the standard principal-agent model are based on a flawed understanding of human behaviour. Standard agency theory must be repaired. Behavioural agency theory, as described in Chap. 5, provides a much better framework for designing rewards and incentives than the standard model.

A THIRD POSSIBLE FUTURE FOR THE PUBLIC CORPORATION

In *The Modern Corporation* Berle and Means describe a third possible future for the public corporation. It is often overlooked.[11] They suggest the possibility of retaining the benefits of public corporations, while at the same time ridding society of the corporation's attendant evils.[12]

> When a convincing system of community obligations is worked out and is generally accepted, in that moment the passive property right of today must yield before the larger interests of society. Should the corporate leaders, for example, set forth a program comprising fair wages, security to employees, reasonable service to their public, and stabilization of business, all of which would divert a portion of profits from the owners of passive property, and should the community generally accept such a scheme as a logical and human solution of industrial difficulties, the interests of passive property owners would have to give way. Courts would almost of necessity be forced to recognize the result, justifying it by whatever of the many legal theories they might choose. It is conceivable,- indeed it seems almost essential if the corporate system is to survive,- that the "control" of the great corporations should develop into a purely neutral technocracy, balancing a variety of claims by various groups in the community and assigning to each a portion of the income stream on the basis of public policy rather than private cupidity.[13]

---

[11] See Bratton, W., & Wachter, M. (2010). Tracking Berle's footsteps: the trail of the Modern Corporation's last chapter. *Seattle University Law Review, 33*(4), pp. 849–875.

[12] For the evils that attend public corporations see, for example, the (somewhat polemical) book *The Corporation – The Pathological Pursuit of Profit and Power* by Joel Bakan (2004). For a philosophical argument that corporations are in effect private governments or dictatorships, see Anderson (2017).

[13] Berle, A., & Means, G. (1932) p. 355.

This third option is curiously prescient, anticipating many of the ideas about the repaired theory of executive agency which have been explained in the preceding chapters. These various ideas are summarised in the ten propositions set out below which are put forward in the spirit of Berle and Mean's third potential future for the public corporation.

Proposition 1
The standard model of executive agency, which has had a major impact on management theory and practice in the last 30 years, is flawed and in urgent need of repair.

Proposition 2
The doctrine of (short-term) shareholder value maximisation, advocated by Milton Friedman and others, is misconceived—it does not serve the best long-term interests of companies, shareholders, employees, or society. It should be replaced by a new doctrine of long-term total firm value maximisation. Directors, investors, employees, and any other important stakeholders should be encouraged to unite around this new doctrine. Long-term total firm value maximisation should become the primary objective of all public corporations. If necessary this principle should be enshrined in company law and financial regulations.[14]

Proposition 3
Public corporations have too much ontological substance to be dismissed as mere legal fictions. Corporations are real entities with identities, temporal existence, corporate cultures, and physical presence. They have legal and ethical responsibilities commensurate with their positions in society.

Proposition 4
Corporate managers have fiduciary responsibilities of a higher legal and ethical standard than those implied by an agency relationship. All senior exec-

---

[14] In the UK this would mean amending section 172 (1) of the Companies Act 2006 to make it clear that directors have a duty to promote the long-term success of the company for the benefit of all major stakeholders. Changes announced by the Department of Business, Energy and Industrial Strategy in June 2018 (The Companies Miscellaneous Reporting Regulations 2018) go some way towards this by requiring directors to report on how they have engaged with a wide set of duties contained in section 172. This requires them to have regard, among other matters, to the interests of employees, supplies, customers, the community and the environment, and to act fairly as between members.

utives should be encouraged to recognise the significance of this high level of ethical responsibility. Companies should report on how directors and senior executives have engaged with their fiduciary responsibilities.

Proposition 5
Company managers should be rewarded for their value-creating activities with generous fixed salaries and modest bonuses. The interests of shareholders and managers should be aligned by requiring executives to invest cash bonuses in company shares or by rewarding them partially with restricted stock. Highly leveraged long-term incentive plans incorporating complex performance conditions are not the answer.

Proposition 6
We should not assume that there is a general problem of executive motivation. The greater risk is that high-powered performance-based incentives will crowd-out intrinsic motivation. The remuneration committee's dilemma cannot be solved by designing more sophisticated incentives. Instead, the dilemma must be "dissolved" rather than "solved"[15] by placing greater focus on intrinsic motivation. We want top executives who are, in the terminology of Julian Le Grand, more "knightly" than "knavish".[16]

Proposition 7
Shareholders own shares, which have rights to dividends, votes, and assets in a winding-up, but they are not in any other meaningful sense the sole "owners" of public corporations. Others, especially employees who have made investments of specific human capital in their employing companies, also have stakeholder participation rights, which should be reflected in governance arrangements.

Proposition 8
There should be broader participation in company governance. Major shareholders should form investor committees modelled on Swedish nom-

---

[15] In the same way that Karl Popper resolved the problem of induction by turning it on its head and focusing on falsifiability rather than verifiability, I am suggesting that the remuneration committee's dilemma can be "dissolved" (in the sense of being "made to go away") by placing more attention on intrinsic, rather than extrinsic, motivation.

[16] Le Grand, J. (2003). *Motivation, Agency and Public Policy*. Oxford: Oxford University Press.

ination committees to advise companies on the appointment of directors and executive pay.[17] Other individuals who are affected by resource allocation rules should have representation rights in company governance systems. These might include works councils, employee advisory panels, and worker representation on company boards or major committees, including the remuneration committee.

Proposition 9
Rewards should be allocated in proportion to inputs, including both capital and labour. Rules that respect proportionality are more likely to be regarded as equitable, whereas rules that dis-proportionately benefit elites will be perceived as unfair. Perceived fair pay is an important characteristic of high-trust organisations. If shareholders, and employees generally, believe that senior executive pay is excessive, then confidence in top management will be undermined. Companies should be encouraged to produce "fair pay" reports.[18]

Proposition 10
Where necessary corporate law should be amended in order to bring about change. Otherwise, companies should be encouraged to devise governance arrangements that are best suited to local conditions. Regulators should enable and support local governance that complies with the law and with these principles.

FINAL WORDS

Professor Simon Deakin of Cambridge University concludes his 2012 paper "The corporation as commons", which I discussed at some length in Chap. 4, as follows: "the sustainability of the corporation depends on ensuring proportionality of benefits and costs with respect to the inputs made to corporate resources, and on the participation of the different stakeholder groups in the formulation of the rules governing the management of those resources".[19] Colin Mayer of Oxford University has issued

---

[17] See Chap. 4, n22.
[18] See, for example, the "fair pay charter" included in Standard Chartered Bank's directors' remuneration report for 2017 (p. 84 of the bank's Annual Report 2017).
[19] Deakin, S. (2012). The corporation as a commons: rethinking property rights, governance and sustainability in the business enterprise. *Queen's Law Journal*, 37 (2), p. 381.

a similar warning in his book *Firm Commitment*.[20] The very future of the public corporation is at stake.

As an epigram to this short book, which has sought to repair agency theory in so far as it applies to shareholders and executives in public corporations, I recall a remark once made by the famous economist Alfred Marshall: "the work I have set before myself is this – how to get rid of the evils of competition while retaining its advantages".[21] Marshall urges other scholars of business and economics to work to similar ends. His concerns about the "evils of competition" apply to capitalism in its entirety. For all its strengths as a wealth production system, unrestrained capitalism has major flaws, as we have found out once again. One such flaw is the remuneration committee's dilemma—the risk of executive pay inflation that is not good for the economy or for society. We must fix this problem.

Further Reading
This chapter has referred to a number of books which are relevant to the future of capitalism and the public corporation. In particular, I would recommend Jacobs, M., & Mazzucato, M. (2016). *Rethinking Capitalism*. Wiley Blackwell, and Mayer, C. (2013). *Firm Commitment: Why the Corporation is Failing Us and How to Restore Trust in it*. Oxford University Press.

REFERENCES

Anderson, E. (2017). *Private Governments – How Employers Rule Our Lives and Why We Don't Talk About It*. Princeton/Oxford: Princeton University Press.
Ashraf, N., Camerer, C., & Loewenstein, G. (2005). Adam Smith, Behavioral Economist. *Journal of Economic Perspectives, 19*(3), 131–145.
Bakan, J. (2004). *The Corporation – The Pathological Pursuit of Profit and Power*. New York: The Free Press.
Berle, A., & Means, G. (1932). *The Modern Corporation and Private Property*. New York: Macmillan.

---

[20] Mayer, C. (2013). *Firm Commitment: Why the Corporation is Failing Us and How to Restore Trust in it*. Oxford: Oxford University Press.

[21] Quoted by Keynes, J. (1956) "Alfred Marshall, 1842–1924". In A. Pigou (Ed.), *Memorials of Alfred Marshall*. New York: Kelley & Millman, Inc, p. 16.

Jacobs, M., & Mazzucato, M. (2016). *Rethinking Capitalism*. Chichester: Wiley Blackwell.

Kay, J. (2003). *The Truth About Markets: Their Genius, Their Limits, Their Follies*. London: Allen Lane.

Keynes, J. (1956). Alfred Marshall, 1842–1924. In A. Pigou (Ed.), *Memorials of Alfred Marshall*. New York: Kelley & Millman, Inc.

Mason, P. (2015). *Post Capitalism: A Guide to Our Future*. London: Allen Lane.

Mazzucato, M. (2018). *The Value of Everything: Making and Taking in the Global Economy*. London: Allen Lane.

Norman, J. (2018). *Adam Smith: What He Thought and Why It Matters*. London: Allen Lane.

Streeck, W. (2016). *How Will Capitalism End? Essays on a Failing System*. London: Verso.

INDEX[1]

[1] Note: Page numbers followed by 'n' refer to notes.

© The Author(s) 2019
A. Pepper, *Agency Theory and Executive Pay*,
https://doi.org/10.1007/978-3-319-99969-2

CPI Antony Rowe
Eastbourne, UK
February 08, 2019